Kick & Code

How AI is Revolutionizing Soccer

By
Jordan Blake

Kick & Code

How AI is Revolutionizing Soccer

Table of Contents

Introduction

Soccer, often referred to as the "beautiful game," has always been about passion, skill, and strategy. But as we move deeper into the 21st century, the intersection of technology and soccer is becoming not just a talking point but a transformative force. From analyzing player movements to enhancing fan engagement, artificial intelligence (AI) is carving out a significant role in the sport. This book aims to shed light on how AI is revolutionizing soccer, offering readers a comprehensive view of its far-reaching impact.

Over the past decade, advancements in technology have revolutionized many aspects of our lives, and soccer is no exception. Once a game purely driven by instincts and raw talent, it has evolved into a sophisticated sport where data and technology play crucial roles. Think about it: every dribble, pass, and shot can now be tracked, analyzed, and optimized. That's the power of AI in action.

Artificial intelligence is more than just a buzzword – it's a catalyst for change. Its applications in soccer are multifaceted, touching everything from game strategy and player performance to fan experience and beyond. The introduction of wearable technology, real-time data analytics, and predictive algorithms has created a new playing field both on and off the pitch. For coaches, analysts, and players, AI is offering unparalleled insights into the game's intricacies, transforming traditional methods into modern marvels.

One might wonder why AI is so crucial in soccer. To put it simply, it's about making better decisions faster. In a sport where a split-second can make the difference between victory and defeat, having access to real-time, data-driven insights can be the ultimate game-changer. It's about elucidating patterns and trends that human eyes might miss, allowing for a deeper understanding of the game's dynamics and nuances.

Take game performance enhancement, for instance. With AI, real-time data analysis can help coaches make tactical adjustments during a match, providing a competitive edge. Wearable technologies monitor players' physiological conditions, predicting injuries before they happen and optimizing training loads. The scouting and recruitment process, once heavily reliant on human scouts, is now augmented with AI algorithms that identify promising talent with pinpoint accuracy.

The convergence of AI and soccer is not just limited to those on the field. Fans are experiencing a new level of engagement through AI-driven platforms that offer personalized content, interactive experiences, and virtual environments. Broadcasters are utilizing AI to deliver intelligent commentary and real-time statistics, creating a richer viewing experience. The ripple effect of AI's integration into soccer extends to every corner of the sport – making it more exciting, engaging, and innovative than ever before.

However, it's essential to acknowledge that this technological revolution doesn't come without its challenges. Data privacy concerns, the ethical implications of AI, and the potential for over-reliance on technology are significant issues that need careful consideration. Striking a balance between leveraging AI's capabilities and maintaining the sport's human touch is crucial. After all, soccer is not just a game of numbers and algorithms; it's a game of heart, spirit, and human connection.

This book is structured to offer an in-depth exploration of AI's transformative role in soccer. From the early days of soccer analytics to the latest advancements in AI technologies, we will journey through the evolution of how data and technology have shaped the sport. We'll delve into game performance enhancement, player monitoring, scouting, training, and much more, each chapter providing detailed insights and real-world examples.

As we navigate through these topics, we'll also highlight case studies from top clubs, interviews with AI pioneers, and perspectives from coaches and analysts. These stories will illustrate AI's impact on the field and provide practical takeaways for those looking to integrate AI into their own soccer environments. Whether you are a coach aiming to refine your team's strategy, a player seeking to enhance your performance, or a fan eager to understand the game's future, this book offers something for everyone.

In the realm of youth development, AI is paving the way for identifying talent early and nurturing it through enhanced training programs. The financial implications of AI also present a fascinating area of exploration, from revenue generation to cost-benefit analyses. And let's not forget the growing influence of AI in women's soccer, breaking barriers and setting new standards.

Looking ahead, the future trends of AI in soccer promise even more innovative breakthroughs. Emerging technologies and predictive insights will continue to redefine the way the game is played and experienced. Yet, as with any technological advancement, it's vital to remain vigilant about the risks and limitations. Over-reliance on technology, addressing AI bias, and maintaining a balanced view are topics that require ongoing dialogue and critical assessment.

Ultimately, this book aims to provide a balanced view of AI's role in soccer. While we celebrate the incredible advancements and opportunities AI brings, we also remain mindful of preserving the game's essence. Soccer, at its core, is a sport driven by human passion, creativity, and unpredictability. As we embrace AI's transformative power, we must ensure that the "beautiful game" retains its soul.

As you embark on this journey through the pages ahead, we hope you gain a deeper appreciation of how AI is shaping the future of soccer. We invite you to explore the possibilities, challenges, and innovations that lie at the intersection of technology and sport. Let's embark on this exciting journey together, discovering the myriad ways AI is bringing a new dimension to the world's most beloved game.

Chapter 1:
The Evolution of Soccer Analytics

The journey of soccer analytics has been nothing short of transformative, transcending the humble beginnings of scribbled notes and rudimentary statistics. As the sport evolved, so did the methods of analyzing play, shifting from basic match reports to nuanced data interpretation. With the rise of technology, data-driven decision making emerged, revolutionizing how teams strategize and perform. Now, richly detailed metrics allow coaches to fine-tune tactics with unprecedented precision, and players benefit from targeted training regimens informed by complex algorithms. This era of analytics has not only changed how the game is played but also deepened our understanding and appreciation of its intricate beauty. Soccer is no longer just a game of passion and skill—it's a blend of art and science, driven by numbers and bolstered by innovation.

Early Days of Soccer Analysis

Before data dashboards and advanced metrics became the cornerstone of soccer strategy, the sport relied heavily on the intuition and expertise of coaches, scouts, and analysts. The early days of soccer analysis were a distinct period characterized by rudimentary tools and manual calculations. The journey started with coaches annotating plays on chalkboards and scouts taking handwritten notes during matches. This traditional approach, while constrained by limited technology, laid the groundwork for what would eventually become a data-centric revolution.

In the mid-20th century, statistical analysis began to infiltrate soccer, albeit very slowly. Analysts like Charles Reep of England pioneered the collection of match data. Reep meticulously recorded pass completions, shots, and other key events on paper spreadsheets. He aimed to decode the game's complexities into quantifiable insights. Although his work was groundbreaking, it also courted controversy. Critics argued that his findings oversimplified the game and did not account for its fluid complexity. Despite this, Reep's contributions were pivotal, igniting curiosity and debate that would drive the quest for deeper understanding.

Reep's approach focused heavily on the principle that maximizing scoring opportunities could be achieved by favoring fewer, longer passes. While his ideas had their merits, they also demonstrated the limitations of early soccer analysis. With no sophisticated tools to capture the nuances of play, analysts had to rely on basic statistics that could sometimes mislead. Teams attracted to Reep's high-risk, high-reward strategies often ignored the subtleties that define successful soccer, such as player positioning, movement off the ball, and teamwork. Nevertheless, Reep's work represented a significant shift from intuition-based to data-driven decision-making, laying the first bricks in the foundation of soccer analytics.

Simultaneously, across the Atlantic, the rise of sabermetrics in baseball, popularized by Bill James, started to exert its influence on soccer. Although the sports differ greatly, the principles behind sabermetrics – a systematic, empirical approach to understanding performance – transcended disciplines. The concept of breaking down a sport into individual, measurable components and then using those insights to make strategic decisions was alluring. A small but growing community of soccer analysts began embracing this new paradigm, understanding the game not just as an art, but also as a science.

These early pioneers often worked in isolation, facing skepticism from traditionalists who believed that the beautiful game should not be reduced to numbers and charts. Analysts often had to battle against entrenched mindsets that valued experience and gut feeling over empirical evidence. It was an uphill climb that required not just intelligence and dedication, but also a deep love for the sport and a belief in the potential of data to unlock new dimensions of the game.

Technological limitations of the time also hampered the speed and scope of early soccer analysis. The absence of computers for real-time data processing meant that analysis was labor-intensive and time-consuming. Analysts spent hours manually logging data from match recordings, and the insights gained were often available only well after the fact. While tedious, this labor-intensive process did train a generation of analysts to deeply understand the game, as they had to watch and rewatch footage, learning to discern patterns and anomalies.

By the late 1980s and early 1990s, the advent of personal computers and rudimentary statistical software began to empower a new wave of soccer analysts. Programs like Microsoft Excel allowed analysts to perform more complex statistical calculations and graphical representations of data. The potential for data to provide strategic insights was becoming clearer. Though still primitive by today's standards, these tools represented significant advancements and democratized access to analytical techniques that were previously the preserve of those with specialized skills.

A notable milestone in the evolution of soccer analytics was the introduction of OPTA in the mid-1990s. OPTA began to collect and categorize detailed match data, making it accessible for clubs, media, and analysts. Its database included a wealth of information about player actions, such as passes, tackles, and shots, providing a richer, more granular view of the game. With this access, analysts could begin to study soccer in greater depth, applying statistical models that were earlier used in other sports.

The availability of OPTA data heralded the beginning of the modern era of soccer analysis. Clubs could now cross-reference performance data with video analysis, yielding more actionable insights. Analysts could engage in comparative studies, benchmarking players against each other and historical data, thus elevating both scouting and match preparation. This period saw an uptick in the professionalization of soccer analysts, as clubs and organizations increasingly recognized the value of objective, data-driven insights.

In parallel, academic interest in soccer analytics grew. Researchers began to publish papers examining various aspects of the game, from spatial analysis to player efficiency metrics. Journals dedicated to sports science and analytics provided a platform for these studies, thereby legitimizing and solidifying soccer analytics as a field worth serious academic and practical investment. Partnerships between academic institutions and soccer clubs began to emerge, further advancing the discipline.

As the 21st century dawned, the immersion of the tech industry into sports analytics began to take hold. Companies specializing in data capture and analysis started offering services tailored to soccer. The fusion of sports and technology accelerated the development of more refined metrics and models. By harnessing advanced algorithms and improved data processing capabilities, soccer analysis transitioned from a niche interest to a burgeoning industry, poised to transform every level of the sport.

Thus, the early days of soccer analysis were a fertile ground, seeded with exploration and curiosity. The contributions of these early analysts, though sometimes contentious and rudimentary, cannot be overstated. They opened the door to a new understanding of the game, treating it as a complex system that can be dissected, examined, and optimized. In doing so, they paved the way for the sophisticated, AI-driven analytics that we see transforming soccer today.

Rise of Data-Driven Decision Making

The evolution of soccer analytics has fundamentally reshaped the sport, with one of the most pivotal developments being the rise of data-driven decision making. As teams began to understand the sheer volume of information available through player tracking and event data, the groundwork was laid for a more analytical approach to the game. This shift, which started gaining momentum in the early 2000s, was built upon the simple realization that decisions grounded in data can significantly enhance performance and outcomes.

In this data-rich environment, decision making has transitioned from intuition-based choices to those supported by robust statistical analysis. Coaches and analysts now have access to detailed metrics that cover everything from player movements to ball possession patterns. These insights allow for a more nuanced understanding of both strengths and weaknesses, offering a competitive edge that was unimaginable in the earlier days of soccer analysis.

One of the most significant impacts of data-driven decision making is in tactical planning. By leveraging data, teams can analyze the playing styles of their opponents with unprecedented detail. They can predict likely formations, identify key players who can turn the game, and develop strategies specifically designed to counter the opposition's strengths while exploiting their weaknesses. This level of preparedness was unheard of before the era of data analytics.

For instance, consider set pieces, an area where data analytics has made enormous strides. Detailed analysis of past games reveals patterns in how teams defend and attack during corners or free kicks. This data is then used to design rehearsed routines that can catch opponents off guard, maximize scoring opportunities, and minimize risks. Consequently, the once static and unpredictable elements of the game have become more controllable and measurable.

Additionally, player performance is continually scrutinized through the prism of data. Metrics like expected goals (xG), passes completed, and sprint distances provide a more objective measure of player contributions rather than relying solely on traditional statistics such as goals and assists. This has transformed talent evaluation, making it possible to uncover undervalued players who excel in specific areas critical to a team's success.

Moreover, the granularity of data allows for more personalized coaching. With wearable technology and GPS tracking, every player's movements can be monitored in real-time. Coaches utilize this data to tailor training programs to address individual weaknesses and enhance strengths. Players receive precise feedback on aspects like their positioning, movement efficiency, and even their mental and physical fatigue levels.

Breakthroughs in machine learning and artificial intelligence have only amplified the potential of data-driven decision making. AI models can now process vast datasets quickly, identifying patterns and trends that might elude even the most experienced human analysts. By predicting outcomes based on historical data, these models offer actionable insights that can be decisive in match preparations and in-game adjustments.

Beyond the pitch, clubs have realized the financial benefits of data-driven decision making as well. From scouting and recruitment to contract negotiations, data provides a clearer picture of a player's value. This is particularly useful when signing new talent or renegotiating existing contracts. Instead of relying on subjective assessments, clubs can present quantifiable reasons for contract terms, making the process more transparent and fair for both parties involved.

Furthermore, fan engagement has been revolutionized through data dissemination. Real-time statistics and detailed post-match analyses have captivated fans and provided them with deeper insights into the game they love. This not only enhances their viewing experience but also fosters a more informed and engaged community. Broadcasters and commentators now employ advanced metrics to explain the science behind what happens on the pitch, making the game more accessible and exciting to fans.

It's worth mentioning the psychological aspect of data-driven decision making. Having a factual basis for decision making reduces the burden of uncertainty and the pressure of making gut-based choices. This shift can result in a more composed and focused team environment where decisions are respected and trusted because they are backed by empirical evidence.

In summary, the rise of data-driven decision making has been a game-changer in the evolution of soccer analytics. The ability to make informed decisions based on comprehensive data has enhanced tactical approaches, improved player performance, and provided new avenues for financial gains. As data collection and analysis technologies continue to evolve, so too will the depth and breadth of insights available, further solidifying data's indispensable role in soccer. The game has not just kept pace with the technological revolution but has embraced it, setting the stage for an even more analytical and strategic future.

Chapter 2:
AI Fundamentals: A Primer
for Soccer Enthusiasts

As soccer continues to evolve, artificial intelligence (AI) stands as a transformative force reshaping the beautiful game. But what exactly is AI? In essence, it's a branch of computer science focused on creating systems capable of performing tasks that typically require human intelligence, such as learning, reasoning, and problem-solving. For soccer enthusiasts, understanding the basic tenets of AI is crucial to appreciating its impact on the pitch. Key AI technologies—ranging from machine learning and neural networks to computer vision—are starting to play pivotal roles in interpreting game data, enhancing training methods, and revolutionizing fan engagement. This primer aims to demystify AI, laying a foundational understanding before delving into its specific applications in soccer. By grasping these fundamentals, you'll gain a deeper insight into how algorithms and data analytics are not just buzzwords but are transforming how the game is played, analyzed, and enjoyed.

What is AI?

Artificial Intelligence (AI) refers to the development of computer systems capable of performing tasks that usually require human intelligence. These tasks can range from speech recognition and decision-making to visual perception and natural language processing. Essentially, AI involves teaching machines to learn from experience, adapt to new information, and execute actions in a way that mimics human cognitive functions.

To better grasp what AI entails, picture a soccer coach who can process countless data points during a match: player positions, game tempo, opponent strategies, and even weather conditions. Now, imagine if a machine could not only do the same but also predict outcomes, suggest real-time tactical changes, and analyze player performances with unerring precision. That's the promise AI holds for soccer enthusiasts.

AI isn't a monolithic concept but rather an umbrella term that encompasses a variety of technologies and approaches. Machine learning, a subset of AI, allows systems to automatically improve their performance through experience—learning from past data to make better predictions. Natural language processing enables machines to understand and interact using human language, making communication seamless. Computer vision technology allows machines to interpret and make decisions based on visual data such as video footage.

Machine learning algorithms, for instance, can identify patterns in massive datasets, making it possible to predict a player's future performance based on historical game data. Similarly, computer vision can analyze hours of match footage to provide insights that even the most experienced human analyst might miss. These capabilities are invaluable in the high-stakes world of soccer, where split-second decisions can determine the outcome of a match.

Understanding AI also means recognizing its evolutionary nature. Early systems required explicit programming to perform tasks. However, modern AI systems, thanks to machine learning and deep learning, can autonomously improve without human intervention. This leap has far-reaching implications for soccer, where continuous adaptation and learning are crucial for success.

Complexity doesn't mean inaccessibility. At its core, AI aims to simplify and optimize decision-making processes, making sophisticated analyses more accessible to managers, coaches, and even fans. For soccer enthusiasts, this means a transformative experience—one where data-driven insights enhance every aspect of the game.

In the soccer domain, AI's primary functions can be divided into reactive machines, limited memory, theory of mind, and self-aware AI. Reactive machines, like those used in many current AI applications, respond to specific inputs with pre-programmed responses. Limited memory systems can use past experiences to inform current decisions—a critical feature for AI used in tactical game analysis. As the technology advances, soccer will benefit from more sophisticated AI systems capable of understanding and predicting human emotions and mental states, factors crucial in high-pressure games.

By understanding what AI entails, we appreciate its potential to revolutionize soccer. It's not just about automation or replacing human roles but augmenting human capabilities. With AI, a coach doesn't merely instruct players; they gain a partner that can offer nuanced, data-driven insights. A fan doesn't just watch a game; they experience an enhanced narrative, rich with analytical depth.

The ripple effect extends beyond the pitch. In scouting and recruitment, AI can sift through countless player profiles to identify hidden gems, predict future stars, and even assist in negotiating contracts with unparalleled precision. Wearable technology, enhanced by AI, can monitor player health, reduce injury risks, and extend careers, while virtual coaching assistants offer personalized training regimens.

Moreover, AI in soccer creates a more inclusive environment. Its accessibility democratizes high-level insights and analytics, making them available to amateur teams, smaller clubs, and even grassroots initiatives. It levels the playing field, allowing talent and strategy to shine irrespective of budget constraints. Imagine a world where a small club in a lower division uses AI-driven insights to defeat a top-tier team. That's the future we're speeding towards.

A key characteristic of AI is its continual learning and adaptation. Just as a player hones skills through practice and matches, AI systems refine their algorithms through data. This means that the AI systems we see today in soccer are only scratching the surface of their potential. As more data is fed into these systems, their predictions, analysis, and strategies will become even more accurate and tailored.

Such transformation is underscored by the integration of machine learning and deep learning techniques, forming the backbone of AI applications in soccer. Machine learning models analyze past game data to uncover hidden patterns and relationships, providing actionable insights into team performance, opponent strategies, and individual player metrics.

Neural networks, a type of deep learning model, enable AI systems to simulate human brain function, allowing for more sophisticated data analysis. These networks can process vast amounts of information from various sources—match footage, player statistics, biometric data—to generate comprehensive reports that help coaches and analysts make informed decisions.

As AI becomes more embedded in soccer, it's crucial to consider its broader implications. Ethical considerations, data privacy, and the balance between technology and human touch will shape its future trajectory. Yet, one thing remains clear: AI stands to redefine the beautiful game in ways we're only beginning to understand.

The promise of AI in soccer isn't just in its technical capabilities but in its potential to inspire new ways of thinking about the game. It challenges traditional approaches and opens the door to innovation, creativity, and a deeper understanding of soccer's intricate dynamics. It invites everyone, from coaches to fans, to imagine what's possible when human passion intersects with technological prowess.

As we delve deeper into AI's applications in this book, it's essential to appreciate its transformative potential. AI doesn't merely fit into soccer's existing framework—it reshapes it, offering new opportunities and redefining old challenges. It's an exciting evolution, promising a future where the love for the game is enhanced by the brilliance of artificial intelligence.

Key AI Technologies Impacting Soccer

Artificial intelligence is revolutionizing soccer in ways we couldn't have imagined just a decade ago. It's not just about automating tasks or generating insights that would take humans days to collate; AI is shaping the game's very fabric, and its influence is felt on multiple fronts. From performance analytics and tactical adjustments to player monitoring and fan engagement, AI is becoming an indispensable tool in modern soccer.

One of the most game-changing technologies in soccer is real-time data analysis. Advanced algorithms can process data from myriad sources—tracking systems, video feeds, wearables—in milliseconds. This data crunching enables coaches to make minute-by-minute adjustments based on actual game conditions rather than gut feeling or historical trends. Imagine a scenario where a coach knows the exact moment a player is showing signs of fatigue or a slight drop in form, allowing for a timely substitution that could be the difference between a win and a loss.

Another key area where AI is making waves is in tactical decision-making. AI-driven software can analyze an opposing team's strategies by studying hours of game footage and identifying patterns that would be impossible for a human analyst to discern. These powerful tools can provide recommendations for optimal defensive formations or pinpoint weak spots in the opponent's defense, essentially serving as a virtual assistant coach working tirelessly in the background.

The realm of player performance monitoring has also been profoundly impacted by AI. Wearable technologies equipped with sensors can track everything from heart rates and running speeds to muscle activity and stress levels. This data can then be fed into AI algorithms designed to predict potential injuries before they occur. By identifying signs of strain or fatigue that even the most perceptive coach might miss, AI can help extend players' careers and keep them on the pitch longer and in better shape.

AI is also transforming the scouting and recruitment process. Using machine learning and predictive analytics, clubs can now sift through vast amounts of data to identify young talents who might otherwise go unnoticed. Algorithms can evaluate a player's potential by analyzing various performance metrics, offering a more comprehensive picture than traditional scouting reports could provide.

In training, AI provides a personalized touch at a scale previously unimaginable. Virtual coaching assistants use data from individual players to create customized training programs that focus on their specific needs. These programs can be adjusted in real-time based on ongoing performance data, ensuring that training regimens are as effective as possible. This level of personalization ensures that each player gets the tailored guidance and support they need to develop to their full potential.

The impact of AI isn't limited to on-field performance alone; it extends to fan engagement and experience. AI-driven chatbots and virtual fan environments provide a more interactive and personalized experience for supporters. These chatbots can answer questions, provide statistics, or even simulate conversations with fans, creating a deeper bond between the club and its followers.

Moreover, broadcasters have begun using AI to enhance live match experiences. Intelligent commentary systems can analyze ongoing plays and provide richer, data-driven insights in real time. Real-time statistics offered during broadcasts allow viewers to engage more deeply with the game, making the entire experience much more informative and enjoyable.

Of course, with the integration of AI comes the challenge of data privacy and ethical considerations. The vast amount of data being collected and analyzed must be handled with utmost care to protect the privacy of individuals involved. Clubs and organizations need to establish protocols that ensure data is used responsibly and ethically.

Ultimately, AI's role in soccer is to augment human capabilities, providing tools and insights that can elevate the game to new heights. By reducing the load on human analysts and offering precise, data-driven insights, AI allows coaches, players, and even fans to focus on what they do best—enjoying the beautiful game.

As AI technologies continue to evolve, their applications in soccer will only expand. We've already seen what's possible with current technology, but the future holds even more promise. Emerging technologies like deep learning and neural networks could push the boundaries of what AI can do, offering deeper insights and more sophisticated solutions.

Clubs that embrace these technologies and integrate them into their operations will likely find themselves at a competitive advantage. With AI's help, they can make more informed decisions, achieve better on-field results, and create a more engaging experience for their fans. This transformation isn't just a trend; it's a fundamental shift in how the game is played, managed, and enjoyed.

So, what's next? The adoption of AI in grassroots soccer, community initiatives, and even the global stages of the game will likely continue to grow. As AI technologies become more accessible, their benefits will permeate all levels of the sport, democratizing the advantages they offer and bringing new opportunities for players and fans alike.

For now, what's clear is that AI has the power to transform soccer profoundly. From strategy and scouting to training and fan engagement, AI technologies are reshaping every aspect of the game. Embrace these changes, and we'll pave the way for a future where soccer is not just a sport but a testament to synergy between human ingenuity and machine precision.

Chapter 3:
Game Performance
Enhancement Through AI

Artificial intelligence is revolutionizing soccer by delivering real-time insights that enable teams to optimize their performance dynamically. Whether it's analyzing in-game data to make instant tactical adjustments or using predictive algorithms to foresee the opposition's next move, AI empowers coaches to make smarter decisions on the fly. This transformative technology doesn't just crunch numbers; it translates raw data into actionable intelligence, providing a competitive edge. By integrating machine learning with traditional coaching techniques, soccer teams can now refine their strategies with unprecedented precision and speed, making every game a sophisticated dance of data-driven maneuvers. As AI continues to evolve, its influence on game performance will only deepen, pushing the boundaries of what's possible on the pitch.

Real-Time Data Analysis

In the high-stakes world of soccer, decisions need to be made in the blink of an eye. Real-time data analysis has revolutionized how teams approach strategy and performance within the duration of a match. This approach isn't just about crunching numbers; it's about translating those numbers into actionable insights that can make or break a game. The integration of real-time data is akin to having a tactician whispering invaluable advice straight into the coach's earpiece.

Consider a scenario where a team's top striker is suddenly sidelined with an injury. Traditional methods would leave coaches scrambling to revise their strategies on the fly. With AI-driven real-time data analysis, solutions are not only swift but also backed by solid data. Substitute options and tactical adjustments are highlighted, complete with projections of their potential effectiveness based on historical and real-time data. This eliminates much of the guesswork that plagues even the most seasoned coaches during high-pressure moments.

At the heart of real-time data analysis are sophisticated algorithms that continuously process streams of data collected from multiple sources—player movements, ball trajectory, crowd patterns, and more. Advanced sensor technologies located around the stadium, on players' bodies, and even on the ball feed this data into AI systems for immediate processing. The speed and precision of these systems allow for in-game adjustments that can significantly alter the course of a match.

Let's say the opponent's left back has been identified as a weak link. Real-time data, including fatigue levels and recent performance metrics, suggests that this player is struggling to keep up. Coaches can receive instantaneous alerts and tweak their tactics to exploit this vulnerability, either by instructing wingers to concentrate attacks down that flank or by adjusting the formation to overload that side.

One emblemic example of real-time data at work was during the 2018 FIFA World Cup. Teams like Germany employed real-time analytical tools to monitor both their players and their opposition continuously. As the match unfolded, coaches received constant updates, allowing them to understand the flow of the game better and make critical adjustments.

The benefits go beyond tactical adjustments. For players themselves, real-time data can offer immediate feedback on their performance. Wearable technology relays information regarding speed, heart rate, distance covered, and even impact forces from tackles. This level of granularity helps players to adjust their exertion levels instantly, minimizing the risk of injury and optimizing performance.

Real-time data isn't just useful for coaches and players; it also enhances the fan experience. Stadiums equipped with advanced AI systems can offer spectators an intricate understanding of what unfolds on the pitch. Apps can provide live heat maps, player statistics, and predictive outcomes, making the viewing experience more immersive and interactive. Fans no longer just witness goals; they witness the data-driven tactics behind those goals.

To enable this depth of analysis, partnerships between tech companies and soccer clubs are increasingly common. Companies like Opta, STATSports, and Catapult offer comprehensive solutions tailored to the rigorous demands of professional soccer. These partnerships ensure that clubs have access to cutting-edge AI technologies, which are continually refined and updated to keep pace with the evolving sport.

Another transformative aspect of real-time data is its capacity to assist refereeing decisions. Although we'll delve deeper into refereeing technology in later chapters, it's worth noting that real-time data can significantly reduce human error. AI systems can flag potential offsides, fouls, and other infractions in real-time, providing referees with an additional layer of scrutiny to make more accurate calls.

Grassroots and amateur leagues also stand to benefit from these innovations, making advanced strategies accessible at all levels. Lower-tier teams equipped with real-time data analysis tools can punch above their weight, effectively competing against technologically superior clubs. With community initiatives and more affordable technology, even the smallest of clubs can now harness the same analytical power that was once the preserve of elite teams.

One might ask, how does all this data get processed so quickly? The underlying technology relies on cloud computing and edge computing. While cloud computing deals with the bulk of heavy data processing, edge computing ensures that crucial data is processed near the source, allowing for instantaneous feedback. This hybrid approach ensures that neither speed nor accuracy is compromised, making real-time data analysis a game-changer.

In essence, real-time data analysis serves as the invisible hand guiding a team through a match. While the players and coaches are the visible actors, behind the scenes, AI tirelessly processes data, anticipates scenarios, and proposes optimal strategies. It provides an informed foundation upon which split-second decisions are made, blending the art of soccer with the precision of science, making each match not merely a game but a calculated chess match played on a grand stage.

The transformative impact of real-time data marks the dawn of a new era in soccer. While the core essence of the game—the passion, the unpredictability, the sheer joy of a last-minute goal—remains untouched, the way the game is played and perceived has fundamentally evolved. Real-time data bridges the gap between instinct and informed decision-making, leveling the playing field while elevating the stakes. Soccer, with its rich tradition, has wholeheartedly embraced AI, ensuring the sport continues to captivate and inspire, now fueled by the power of real-time analysis.

tactical adjustments using ai

Artificial intelligence has revolutionized tactical management in soccer, bringing precision and insight to real-time decision-making that was scarcely imaginable a decade ago. In previous eras, coaches heavily relied on intuition and experience to make game-changing decisions.

Today, AI's role in making those crucial tactical adjustments has both demystified and enhanced the art of strategy. Utilizing machine learning algorithms, real-time data, and advanced analytics, AI allows coaches to adapt and refine their strategies on the fly, empowering teams to exploit their opponents' weaknesses and bolster their own strengths.

The real magic of AI in tactical adjustments starts even before the match kicks off. Pre-game analysis tools powered by AI analyze past matches, studying patterns and tendencies of both the team's and the opponent's play styles. Such insights guide coaches on how to set up their formations and decide which players would best exploit the opponent's vulnerabilities. For instance, if an opponent struggles to defend counter-attacks, the AI can recommend a strategy that maximizes speed and quick transitions. Thus, teams can approach each game with a strategic blueprint tailored to their opponent's weaknesses.

Another pivotal element of AI-driven tactical adjustments is real-time data analysis. During a match, AI can process and interpret a vast array of data points from player movements, ball trajectories, and even crowd behavior. This continuous stream of information is fed into predictive models that update in real-time, allowing coaches to gain actionable insights instantly. For example, if AI detects that a particular flank is being exposed, it can recommend a formation shift or substitution to stabilize the defense. Such immediate adjustments can be the difference between securing a win or facing a defeat.

However, it's not just about responding to threats; AI also enhances the ability to seize opportunities. Advanced metrics can identify when a player is in top form or if the opponent's defense is momentarily destabilized. These subtle cues might be missed by the human eye but are flagged by AI, suggesting aggressive offensive plays or tactical fouls to disrupt the rhythm. Coaches can use real-time dashboards and interfaces to visualize these insights, making it easier to translate data into actionable strategies on the pitch.

A crucial aspect of tactical adjustments using AI is player-specific recommendations. AI can drill down to individual performance metrics, suggesting specific roles or responsibilities based on real-time performance indicators. Imagine a scenario where a key midfield player shows signs of fatigue; AI can recommend a substitution or a slight tactical tweak to minimize the impact of that player's reduced output. This level of personalized adjustment ensures that all eleven players contribute optimally to the team's objectives.

Moreover, AI can identify and rectify systemic issues within the team's play. By leveraging pattern recognition and anomaly detection, AI can expose inefficiencies such as predictable passing routes or defensive lapses. For instance, if the team's midfield tends to become overly congested under pressure, AI can suggest spreading out players to widen the field and create more passing options. These systemic tweaks can often turn a well-matched contest in favor of the AI-enhanced team, adding layers of sophistication to traditional tactical approaches.

The ability to experiment and iterate rapidly is another gift AI brings to tactical adjustments. Historical data and simulation models enable coaches to run "what-if" scenarios, exploring how different strategies might unfold in real-world conditions. These simulations can test everything from formation changes to set-piece routines, providing a low-risk environment to identify optimal strategies. Coaches can then implement the most promising tactics in real life, with the confidence of data-backed decisions.

AI is also invaluable for post-game analysis, which blends seamlessly into the next round of tactical planning. Video analysis tools equipped with AI can scrutinize matches frame-by-frame, providing a detailed assessment of every tactical decision made. This feedback loop allows coaches to refine their strategies continuously, improving both in-the-moment decision-making and long-term planning. Identifying patterns and learning from past games, teams evolve and adapt, becoming increasingly resilient and tactically versatile.

Of course, the human touch remains critical. AI's role is to supplement, not supplant, the instincts and experiences of skilled coaches. A coach's understanding of player psychology, team morale, and the nuances of human behavior can't be entirely replicated by AI. However, blending human wisdom with AI-driven insights creates a formidable strategy powerhouse. When a coach and an AI work in harmony, the team benefits from the best of both worlds—sharp data-driven tactics and deeply intuitive human leadership.

In high-stakes competitions, where the margin for error is razor-thin, the nuanced tactical adjustments enabled by AI can be transformative. A single well-timed substitution or a minor change in formation, informed by AI, could pivot the outcome of an entire season. As AI continues to evolve, its role in real-time tactical adjustments will only become more sophisticated, offering even more granular insights and predictive power. This ongoing evolution promises a future where split-second decisions are backed by the unparalleled analytical capabilities of AI, pushing the boundaries of what's strategically possible in soccer.

In conclusion, tactical adjustments using AI signify a leap forward in the strategic aspect of soccer. As teams increasingly adopt AI technologies, the game itself is set to become smarter, faster, and more exhilarating. This synergy between human acumen and artificial intelligence is not just modernizing soccer; it's setting new paradigms for how the beautiful game is played and enjoyed, offering a compelling glimpse into the future of sports strategy.

Chapter 4:
Player Performance Monitoring

In the fast-paced world of soccer, keeping an eye on player performance has become not just a necessity but an art form, tailored by artificial intelligence. In recent years, advancements in wearable technology have revolutionized how teams track crucial metrics like player speed, stamina, and even heart rates, allowing coaches and analysts to optimize training and game strategies. Utilizing sophisticated algorithms, AI can predict potential injuries by analyzing data patterns, thereby preventing issues before they arise and prolonging players' careers. This proactive approach ensures that each player performs at their peak, contributing to the overall success of the team. With AI's precise monitoring capabilities, the once chaotic and unpredictable nature of player performance is rapidly becoming a well-charted territory, making it an invaluable asset in modern soccer's toolkit.

Wearable Technology

In the realm of player performance monitoring, wearable technology has emerged as a transformative force that is redefining how teams and coaches evaluate and enhance player capabilities. Modern wearables, ranging from GPS trackers to heart rate monitors, provide an unprecedented depth of data, allowing for a granular analysis of each athlete's performance metrics. This technological evolution is not merely about collecting data, but about harnessing insights that can revolutionize training regimes, game strategies, and injury prevention.

As soon as they step onto the field, players are now equipped with sophisticated devices that measure everything from velocity and distance covered to heart rate variability and muscle fatigue. These devices are embedded in jerseys, worn as wristbands, or even integrated into footwear, capturing real-time data that flows into a central system for immediate analysis. Coaches and analysts no longer rely solely on visual observations or rudimentary statistics; they now have access to an extensive array of performance indicators that allow for more informed decision-making.

This real-time feedback loop has a significant impact on how training sessions are structured and adjusted. For instance, data showing a player's declining performance during high-intensity drills can lead to immediate modifications in the training schedule, preventing overtraining and reducing the risk of injury. It also helps to tailor individual training programs that align with each player's unique physiological profile, maximizing their development and on-field performance.

Moreover, wearable technology is critical in monitoring players' recovery. Post-match or post-training analysis can reveal how effectively an athlete's body is recovering, identifying areas that need more attention or rest. This can lead to more precise recovery protocols, ensuring players are at their optimal physical condition when it matters the most. For example, sleep monitoring wearables can offer insights into the quality of rest players receive, helping to adjust schedules or environments to enhance recovery rates.

From a tactical standpoint, wearables offer an edge by providing real-time data that can be used to make immediate adjustments during matches. Coaches can track player fatigue and adjust substitutions more strategically, ensuring that the team maintains its competitive edge throughout the game. Additionally, data on player positions and movements can be used to refine tactical approaches, identify gaps in the opponent's formation, and exploit weaknesses more effectively.

The integration of biomechanical sensors also revolutionizes injury prevention. These sensors monitor the nuances of player movement, detecting irregularities that could indicate the onset of an injury. By analyzing gait, joint angles, and muscle activity, these devices can predict potential injuries before they manifest, allowing for preemptive measures. This proactive approach not only extends players' careers but also saves clubs significant costs associated with rehabilitation and lost game time.

Wearable technology doesn't just benefit players and coaches but also offers valuable data to sports scientists and medical staff. The detailed physiological data captured informs better nutritional strategies, hydration plans, and conditioning programs. This multidisciplinary approach ensures all aspects of the player's health and performance are optimized, fostering a holistic environment for athlete development.

A key component of this ecosystem is the software that aggregates and analyzes the wearable data. These platforms use advanced algorithms and AI to identify patterns and provide actionable insights. For example, AI can detect signs of overuse injuries by examining variations in running form or changes in acceleration. This level of insight was previously unthinkable and represents a significant advancement in sports science.

However, the widespread use of wearable technology is not without its challenges. Data security and privacy are paramount concerns. The sensitive nature of the data collected requires strict protocols to ensure it is safeguarded against unauthorized access. Players need to trust that their personal information is being handled responsibly, with explicit consent obtained for its use. Thus, the development of wearable technology must be accompanied by robust data governance frameworks.

The future of wearable technology in soccer looks promising, with advancements continually pushing the boundaries of what is possible. Emerging technologies like smart fabrics and nano-sensors are set to take data collection to new heights. Imagine jerseys that can measure biochemical markers through sweat analysis or insoles that provide continuous feedback on foot pressure distribution. These innovations will further enhance our understanding of player performance and health.

Furthermore, the potential for integrating wearables with other AI-driven systems is immense. By combining data from wearables with video analysis and other performance metrics, a more comprehensive picture of player performance can be achieved. This convergence of technologies will likely lead to even more sophisticated and personalized training and recovery programs.

In summary, wearable technology is a cornerstone of modern player performance monitoring, providing critical data that drives informed decisions at all levels of the game. Its benefits are multifaceted, enhancing not only the physical performance and health of players but also the strategic decisions made by coaches and analysts. As technology continues to evolve, its integration into the world of soccer will only deepen, paving the way for a future where every aspect of the game is optimized through data-driven insights.

Injury Prediction and Prevention

In the high-stakes world of professional soccer, players are continuously pushing the boundaries of physical performance. However, with this constant pursuit of excellence comes an increased risk of injury. Injury prediction and prevention have become critical aspects of player performance monitoring, and artificial intelligence (AI) is at the forefront of these advancements.

Wearable technology, such as GPS trackers, heart rate monitors, and inertial measurement units, generates an immense amount of data on a player's physical status. This data, when analyzed by AI algorithms, can provide deep insights into the physiological and biomechanical state of a player. Through machine learning models, patterns that precede injuries can be identified, allowing teams to take preemptive actions to safeguard their players.

Consider the difference this makes. In the past, an injury might come as a sudden, devastating event. A player's muscle might tear mid-game, ending their season in seconds. Now, AI algorithms can analyze variables such as muscle load, fatigue levels, and biomechanical deviations long before these critical points are reached. If a player's metrics begin to show signs of excessive fatigue or abnormal stress on a particular muscle group, coaching staff can intervene—either by adjusting training loads, modifying practice routines, or even resting the player. This proactive approach significantly reduces the incidence of severe injuries.

The impact of AI on injury prevention isn't confined to physical health alone. Psychological stress and mental fatigue are also significant contributors to injuries. Advanced AI systems can analyze social media activity, communication patterns, and other behavioral data to gauge a player's mental state. By integrating this comprehensive view of both physical and mental workloads, clubs can create more holistic wellness programs tailored to individual needs.

Take, for example, the case of hamstring injuries, prevalent among soccer players. Using machine learning, AI models can examine historical data and real-time metrics to identify early warning signs that precede such injuries, including changes in sprint speed, gait, and frequency of high-intensity movements. By pinpointing these indicators early, the likelihood of a hamstring tear can be significantly reduced, allowing the player to maintain peak performance levels without the interruption of time-consuming recovery periods.

Integrating AI-driven insights with traditional medical expertise creates a potent combination for injury prevention. Sports scientists and medical teams work alongside AI systems, refining algorithms based on new data and emergent medical research. This dynamic relationship ensures that AI's predictions grow more accurate and reliable over time, continually enhancing a team's ability to protect their athletes.

Moreover, the benefits of AI-driven injury prediction and prevention extend beyond individual teams. League-wide databases can be established to share anonymized data, fostering collective learning within the soccer community. This collaborative approach can lead to standardized best practices, ensuring even smaller clubs with limited resources can still benefit from AI innovations.

The predictive power of AI also aids in the rehabilitation phase post-injury. AI can create personalized, data-driven recovery plans, monitoring a player's progress and adapting recommended activities to ensure optimal recovery without the risk of re-injury. By continuously assessing performance data during rehab, AI can predict when a player is fully prepared to return to the pitch, balancing the urgency to resume play with the need for complete recovery.

However, the implementation of AI in injury prediction and prevention isn't without its challenges. Data quality and consistency are paramount. Inconsistent or inaccurate data can lead to faulty predictions, causing more harm than good. Therefore, robust systems for data collection, validation, and storage are necessary. Furthermore, there are ethical considerations regarding data privacy and the extent to which a player's personal data can be monitored and analyzed.

Ensuring transparency and securing player consent are critical aspects of ethical AI deployment in this domain.

Injury prediction and prevention through AI also demands a cultural shift within organizations. Coaches, players, and support staff must trust and understand AI's role in safeguarding player health. This trust is built through education and demonstration of AI's efficacy, making it clear that these technologies are allies in the quest for longevity in athletes' careers.

Soccer clubs around the world are already seeing the dividends from early adoption of AI-driven injury prevention systems. For example, Manchester City's use of AI-powered wearable technology has significantly reduced their incidence of muscle injuries, leading to fewer lost days and a fitter, more resilient squad. Similarly, the German national team has leveraged specialized AI algorithms to optimize player load management, contributing to their international success.

The potential for future advancements in AI-based injury prediction and prevention is immense. As machine learning techniques become more sophisticated, and as wearables evolve to capture even more detailed physiological data, the accuracy and scope of injury prediction will only improve. Imagine a future where ankle sensors predict ligament stress, or where AI-driven video analysis preempts movements that could lead to knee injuries. The horizon is broad, and the technology continues to innovate at a rapid pace.

The benefits are not exclusive to elite professional soccer. Youth academies and amateur leagues can also utilize AI-driven insights to foster talent development while minimizing injury risks. This entails a broader dissemination of technology, ensuring accessibility and affordability across all levels of the sport. By nurturing a healthier, more resilient next generation of players, AI is contributing to the longevity and vibrancy of soccer globally.

In conclusion, injury prediction and prevention stand as a testament to AI's transformative impact on soccer. By combining real-time data analytics with advanced machine learning, clubs can foresee potential injuries and intervene before they occur, dramatically enhancing player performance and career longevity.

The fusion of technology and traditional sports medicine heralds a new era where data-driven insights lead to healthier, more capable athletes. If properly implemented, these advancements promise not just to evolve the game but to protect those who play it.

Chapter 5:
Scouting and Recruitment

In the realm of scouting and recruitment, AI is revolutionizing how clubs identify and secure top talent. Gone are the days when scouts would travel far and wide, relying solely on human intuition and experience. Today, automated talent identification systems analyze vast datasets, from match performances to individual player metrics, bringing unprecedented precision and speed to the process. Predictive performance metrics provide insights that allow clubs to forecast a player's future success with remarkable accuracy, transforming recruitment from an art into a science. This technology not only enhances efficiency but also helps uncover hidden gems who might otherwise go unnoticed, democratizing opportunities and raising the competitive bar across all levels of soccer. AI's role in this domain is nothing short of transformative, promising a new era where data-driven strategies become the cornerstone of building winning teams.

Automated Talent Identification

As soccer evolves through the power of artificial intelligence, one of the most transformative strides lies in automated talent identification. Picture a world where raw soccer talent isn't constrained by geography, where every promising player has a shot at being discovered, whether they're playing in a high-profile academy or a makeshift field in a remote village. This isn't a mere dream but a burgeoning reality, courtesy of AI.

Automated talent identification revolves around deploying sophisticated algorithms to evaluate player potential. Gone are the days when scouts relied solely on anecdotal observations and gut feelings. Today, AI systems analyze vast amounts of data, encompassing player movements, biometrics, and game performance metrics, to discern potential stars. Imagine an algorithm that can parse through thousands of hours of gameplay footage to identify a young player's potential based on specific skill sets like dribbling, passing accuracy, and spatial awareness.

The capability of AI to process and analyze video footage stands at the core of this revolution. Advanced video analysis tools equipped with computer vision and machine learning capabilities can now dissect every aspect of a player's performance. Not only can these systems assess a player's physical attributes like speed and stamina, but they also delve into intricate elements like decision-making patterns and tactical intelligence. This is a game-changer for scouts who can now access in-depth profiles of players from leagues and tournaments worldwide seamlessly and efficiently.

However, the power of automated talent identification stretches beyond mere data analysis. The integration of AI into scouting systems facilitates the creation of comprehensive performance profiles. These profiles are crafted by pulling together diverse data points, such as historical performance data, current fitness levels, and even psychological factors. Through predictive modeling, AI can forecast how young players might develop, their rise or decline in form, or even how injury-prone they might be. This allows clubs to make more informed decisions in their talent acquisition strategy.

One of the remarkable benefits of AI-driven talent identification is its impartiality. Traditional scouting is prone to human biases—a scout may prefer a player for subjective reasons, such as physical appearance or personal affinity. Contrarily, AI evaluates based on a uniform set of metrics, ensuring a fair analysis. This way, deserving talent regardless of background or initial visibility gets the opportunity to shine. The transformative power here is profound; it democratizes talent discovery and diversifies the pool of players reaching professional levels.

Moreover, AI technology aids in uncovering "hidden gems"— those players who may not be on the radar of top scouts due to playing in lesser-known leagues or regions. AI systems don't sleep, take breaks, or miss critical moments. They can continuously monitor global matches, capturing instances of excellence that human scouts might overlook. This broadens the scope of recruitment, enabling clubs to build more versatile and dynamic teams.

While the notion of AI scouts might conjure images of robots replacing humans, the reality is more about augmentation rather than replacement. Human scouts still play a pivotal role, particularly in assessing intangible elements like a player's character, cultural fit, and emotional resilience. Combining human intuition with AI-driven insights creates a powerful synergy, enhancing the precision and effectiveness of the scouting process.

Real-world applications of this technology are already making waves. Leading clubs around the globe have started integrating AI-based scouting platforms into their recruitment departments. These platforms offer valuable recommendations and flag potential recruits that might have escaped traditional scouting methods. The marriage of AI and human expertise ensures a holistic talent assessment approach, heightening the chances of identifying truly standout players.

Additionally, the reach of automated talent identification isn't confined to elite clubs. Smaller clubs and academies are also tapping into this technology, providing them a competitive edge in nurturing homegrown talent. For many of these institutions, AI serves as a force multiplier, enabling them to compete with larger, resource-rich clubs.

By minimizing scouting time and costs, AI helps smaller entities direct their resources more effectively, fostering grassroots development and overall sport growth.

Soccer federations also benefit immensely from these advancements. National teams can track the development of their players from youth levels right through to professional stages, selecting for national duties based on rigorous and objective performance metrics. This ensures that a country's best talents are assembled, not just the most high-profile ones. Furthermore, tracking performance over time assists in identifying trends and areas that need intervention or support, optimizing overall team performance.

However, embracing AI in talent identification isn't devoid of challenges. It requires substantial initial investments in technology and training. Building a robust AI infrastructure involves acquiring high-quality data, developing efficient algorithms, and continually refining these systems to adapt to the ever-evolving nature of the game. Maintenance and updates are crucial to ensuring the system's reliability and accuracy.

Another significant concern revolves around data privacy and security. The wealth of data collected and analyzed by AI systems includes sensitive information about players. Safeguarding this data is paramount, necessitating stringent policies and protocols to prevent unauthorized access and misuse. Transparency in how data is used and ensuring compliance with relevant regulations will foster trust among players, clubs, and fans.

One must also navigate the ethical terrain of automated decision-making. While AI offers objectivity, it doesn't negate the importance of human oversight. Ensuring that final recruitment decisions incorporate human judgment helps in retaining the sport's human essence. Balancing this interaction between AI and human decision-makers will be critical in shaping the future of scouting and recruitment.

Innovation in AI-driven talent identification continues to accelerate. With the advent of more advanced algorithms and machine learning techniques, the precision and predictive power of these systems will only enhance. Technologies like neural networks and deep learning promise even more granular insights, potentially revolutionizing how talent is not just identified but nurtured and developed over time.

The road ahead holds exciting possibilities. As AI penetrates deeper into the realm of soccer, the boundaries of talent identification will expand, leveling the playing field and ushering in a new generation of players who might have otherwise remained undiscovered. This democratization and precision in identifying talent echo the true essence of what AI in soccer can achieve – a thrilling, inclusive, and strategically enhanced sport embraced by fans and stakeholders alike.

Predictive Performance Metrics

In the intricate realm of scouting and recruitment, predictive performance metrics have revolutionized how teams identify and assess potential players. These metrics are derived from a vast ocean of data points, offering scouts and analysts a crystal-clear lens through which they can predict future performance with striking accuracy. No longer are decisions based purely on gut feelings and subjective assessments; AI technologies now allow for a methodical, data-driven approach to talent identification.

Predictive performance metrics encompass a broad spectrum of data sources, including match statistics, physiological data, and even psychological profiles. For instance, a player's sprint speed, stamina, and technique can be quantified and compared across a database of existing players, both past and present. This provides a benchmark, making it easier to identify raw talent that might otherwise be overlooked. Furthermore, AI algorithms can weigh these factors to generate a composite score or ranking, highlighting players who exhibit the most potential.

One standout example of predictive performance metrics is the application of machine learning algorithms that leverage historical data to draw patterns and predict future trends. These algorithms analyze massive datasets, ranging from player stats and game footage to training session performances and even social media activity. By processing these datasets, AI can reveal insights that would otherwise remain hidden, such as a player's likelihood of excelling in particular tactical systems or their potential to develop essential skills over time.

A significant advantage of these metrics is their ability to factor in a myriad of variables simultaneously, offering a multi-dimensional view of a player's capabilities. Traditional scouting methods might rate a forward high for his goal-scoring prowess, but predictive performance metrics can delve deeper. They can analyze the player's off-the-ball movement, his involvement in defensive actions, and even his psychological resilience during high-pressure situations. This holistic view enables teams to make more informed recruitment decisions.

Moreover, predictive performance metrics transcend beyond individual player assessment. They help clubs build a cohesive team by identifying how well a potential recruit might synergize with existing team members. Player compatibility is crucial; AI-driven metrics can assess a player's playing style and mentality, ensuring that new signings will not only fit into the team tactically but also culturally. This creates a harmonious team environment where each player amplifies the strengths of their peers.

Teams also utilize predictive metrics to anticipate a player's development trajectory. By analyzing historical performance trends and growth patterns, AI can predict how a young player might evolve over the upcoming seasons. This serves as a vital tool for clubs looking to invest in youth talent. For instance, a teenage midfielder with excellent ball control and vision but average physical attributes might be predicted to develop significant endurance and strength with proper training. This foresight aids clubs in making strategic long-term investments in players.

Fiscal decisions often hinge on these performance insights. Predictive metrics can influence transfer market strategies by highlighting undervalued talent, thus ensuring economical spending. Conversely, they can also confirm the feasibility of high-profile acquisitions by substantiating the player's prospective value to the team. This duality helps clubs balance their budgets while maintaining competitive squads.

Injuries are another area where predictive performance metrics offer substantial benefits. By incorporating wearables and physiological monitoring, AI can identify patterns indicating an increased risk of injury. For instance, metrics might reveal if a player's muscle fatigue levels are peaking, suggesting a need for rest or adjusted training loads to prevent injury. This proactive approach not only safeguards the player's health but also protects the club's investment, minimizing downtime and maintaining peak performance levels throughout the season.

Furthermore, predictive metrics extend beyond recruitment, influencing coaching strategies and player development programs. AI-generated insights can tailor training regimens to address identified weaknesses or enhance specific strengths, creating a pathway for continuous improvement. This individualized training approach ensures players reach their maximal potential, benefiting both the individual and the team.

The psychological profile of a player is another critical aspect influenced by predictive performance metrics. By analyzing behavioral data, AI can offer insights into a player's mental resilience, confidence, and even leadership potential. For example, a player's reaction to high-pressure scenarios can be quantitatively evaluated, aiding in understanding their mindset and ability to perform under stress. Consequently, clubs can prioritize acquiring not just technically skilled players but those who possess the mental fortitude essential for consistently high-performance levels.

Thus, predictive performance metrics are indispensable in modern soccer scouting and recruitment. They offer an empirical basis for evaluating talent, reducing the reliance on subjective assessments and enhancing decision-making precision. Clubs that harness these metrics gain a competitive edge, identifying and developing talent with a foresighted approach that was unimaginable a decade ago.

In essence, predictive performance metrics are reshaping the future of player scouting and recruitment, turning it from an art into a science. They provide a balanced blend of technical, physical, and psychological data, ensuring that every recruitment decision is backed by thorough analysis. This paradigm shift ensures that the beautiful game continues to evolve, embracing innovation while staying rooted in the principles that make it truly special. Through these innovations, the quest for soccer excellence has become more exciting, bringing strategic depth and rigor into every signing decision.

The new era of scouting and recruitment is here, driven by predictive performance metrics. While the path forward will undoubtedly present challenges, the potential rewards make this a thrilling journey. As technology continues to evolve, the predictive models will only become more refined, offering a continually improving toolkit for those tasked with shaping the soccer stars of tomorrow. In a game where every detail matters, predictive performance metrics are proving to be the ultimate game-changer.

Chapter 6:
AI in Training and Development

Artificial intelligence is revolutionizing how players train and develop, bringing unprecedented levels of customization and efficiency to the process. With AI-powered virtual coaching assistants, players can now receive real-time feedback on their techniques and skills, allowing for immediate adjustments and improvements. Personalized training programs, meticulously designed by AI algorithms, tailor each session to address the individual needs and strengths of players. This means that training is no longer a one-size-fits-all endeavor but a highly specialized experience that maximizes each player's potential. As a result, the integration of AI in training and development is not just enhancing performance but also minimizing the risk of overtraining and injuries, creating a more holistic and sustainable approach to athlete growth.

Virtual Coaching Assistants

In the rapidly evolving landscape of soccer, virtual coaching assistants have emerged as a game-changer. Leveraging advanced AI technologies, these digital coaches are transforming how training and development are approached. They offer a blend of precision, real-time feedback, and personalized insights that human coaches, no matter how experienced, might find challenging to match consistently.

The inception of virtual coaching assistants isn't a recent phenomenon, but their sophistication has notably escalated in recent years. Initially, these assistants were primarily used for basic monitoring and providing generic feedback. Today, they can analyze a player's performance down to granular levels, identifying minute areas for improvement. By doing so, they not only enhance individual skills but also contribute significantly to team dynamics.

What sets virtual coaching assistants apart is their ability to continuously learn and adapt. Using machine learning algorithms, these systems digest vast amounts of data from games, training sessions, and even player biometrics. Every pass, sprint, and tackle is logged and analyzed. The result? Tailored coaching that's specific to the needs of each player, tackling their weaknesses and honing their strengths.

Take, for instance, a scenario where a player struggles with their positioning on the field. Traditional coaching might involve reviewing video footage and receiving verbal feedback. A virtual coaching assistant, however, can provide instantaneous data-driven insights. Through heatmaps, trajectory analysis, and comparison with optimal positional data, the player can visualize their performance and understand precise adjustments they need to make.

Moreover, these assistants aren't confined to post-game or post-training analysis. The embedded AI in wearables and smart equipment means players receive real-time feedback during their sessions. This immediate feedback loop ensures that corrections are made in the moment, leading to quicker skill assimilation and refinement. It's akin to having a coach always by your side, offering guidance with each step you take.

Another key aspect of virtual coaching assistants is their role in injury prevention and recovery. By continuously monitoring players' physical conditions and movements, these systems can flag potential injury risks before they manifest. Patterns indicating overexertion or improper techniques are quickly identified, allowing for timely interventions. This preemptive approach not only prolongs players' careers but also maintains peak performance levels throughout the season.

AI's role in tactical training is equally impressive. By simulating various game scenarios and opponent strategies, virtual coaching assistants prepare players for a wide array of situations. They can recreate the playing style of upcoming opponents, allowing the team to practice and strategize accordingly. These simulations are based on extensive data analysis, making them highly realistic and effective.

Integration with other AI-driven tools further amplifies the benefits of virtual coaching assistants. For example, coupling them with advanced video analysis systems provides a holistic training experience. Players can watch their actions in slow motion, overlay tactical insights, and receive actionable feedback—all within a single platform. This synergistic approach ensures that every element of training is optimized.

The accessibility of virtual coaching assistants is another noteworthy advantage. In grassroots and youth soccer, where resources might be limited, these digital coaches democratize access to top-tier training methodologies. Young athletes in remote or underfunded regions can now receive the same quality of coaching as their counterparts in elite academies. This leveling of the playing field has the potential to uncover and nurture hidden talents from across the globe.

The psychological aspect of training is also catered to by virtual coaching assistants. Stress levels, mental health, and player morale are crucial factors in performance, often overlooked in traditional training setups. Through mood-tracking algorithms and biometric data, these assistants offer insights into a player's mental state, providing recommendations for mental conditioning and relaxation techniques.

Adopting virtual coaching assistants does come with challenges. There's the initial resistance from some coaches and players who might be wary of technology overtaking human expertise. However, the goal isn't to replace human coaches but to augment their capabilities. Human intuition, experience, and motivational skills are irreplaceable, and AI aims to support these elements, creating a more effective and comprehensive coaching environment.

Moreover, continuous updates and advancements in AI technologies ensure that virtual coaching assistants keep evolving. As data collection methods become more advanced and AI algorithms get more sophisticated, the accuracy and effectiveness of these systems will only improve. This evolution promises a future where the integration of virtual and human coaching is seamless and extraordinarily powerful.

To summarize, virtual coaching assistants represent a pivotal shift in soccer's training and development paradigm. By combining rigorous data analysis with real-time feedback and personalized insights, they are revolutionizing how players train, perform, and grow. These AI-driven systems, while still in a burgeoning phase, offer an exciting glimpse into the future of soccer—one where technology and human expertise coalesce to produce world-class athletes and unmatched team performance.

Personalized Training Programs

In the realm of soccer, AI's capacity to drive personalized training programs is nothing short of a game-changer. These programs utilize vast amounts of data to deliver uniquely tailored regimens, revolutionizing how players train and prepare. Long gone are the days when training sessions were one-size-fits-all. Today's AI-driven programs parse through performance metrics, physical health stats, and even psychological factors to create individualized plans that maximize each player's potential.

AI's ability to analyze intricate performance details means that coaches can fine-tune training sessions based on each player's specific needs. For instance, if a player consistently underperforms in endurance drills, AI can suggest adjustments in their cardio routines or recovery periods. Maybe the data indicates a need for increased hydration or nutritional tweaks. These insights go beyond mere conjecture, relying on solid data to pinpoint effective strategies.

Moreover, AI's integration doesn't stop at physical performance. Emotional and mental health parameters are now being considered as well. Algorithms can detect subtle signs of psychological strain, fatigue, or lowered morale, allowing for timely interventions. In a high-stress sport like soccer, maintaining a player's mental well-being is crucial. These tailored interventions can be the difference between a slump and a stellar performance.

Another significant advantage is the use of virtual coaching assistants empowered by AI. These virtual trainers can communicate with players through apps or wearable devices, providing real-time feedback. Consider a scenario where an attacking midfielder needs to improve their passing accuracy. The virtual assistant can guide them through drills specifically designed to enhance this skill, offering constructive critiques immediately following each attempt.

Such customized approaches don't only benefit the players but also liberate coaches to focus more on strategy and less on the minutiae of daily training. By leveraging AI for these detailed insights, coaches can plan more efficiently, ensuring that each session is both productive and purposeful. This optimized use of training time can elevate the entire team's performance, not just that of individual players.

The power of predictive analytics can't be underestimated either. Training programs now employ machine learning algorithms to predict future performance based on current trends. For example, if a player's sprint speed shows a gradual decline over a few months, the system can flag this trend for further investigation. Addressing the issue proactively could involve adjusting their training load or emphasizing recovery, potentially averting a significant dip in performance down the line.

AI's application in personalized training also includes the adaptation of various drills to fit the unique requirements of different playing positions. A goalkeeper's needs differ drastically from those of a forward. AI-driven programs can deliver drills that simulate game scenarios specific to each role, enhancing the relevance and impact of the training.

Furthermore, AI can adapt training intensity based on real-time data, reducing the risk of overtraining and subsequent injuries. By continuously monitoring physical indicators like heart rate, muscle strain, and recovery times, AI can suggest when to dial down the intensity and when a player is ready to push harder. This dynamic adjustment ensures that players are always training in their optimal zones, maximizing growth while minimizing injury risks.

The seamless integration of video analysis tools further augments these personalized programs. Coaches and players can review performance footage, broken down by AI to highlight strengths and areas for improvement. Imagine a defender who needs to improve their positioning during set pieces. By analyzing successful and unsuccessful events, AI can help formulate specific drills to address these gaps.

AI's contributions to personalized training also extend to developing long-term career trajectories. Players' skills and physical attributes evolve, and so should their training routines. AI continually updates these programs based on new data feeds, ensuring that training evolves in sync with the player's development. This adaptability allows for a growth-oriented approach, supporting players at various stages of their careers, from rookies to seasoned professionals.

Customization isn't limited to high-end clubs or elite players. Youth academies and grassroots programs are also benefiting from AI-driven personalized training. These young talents, often overlooked due to generic training methods, now receive attention tailored to their budding skills and potential, fostering early development and increasing their chances of success.

The democratization of AI in soccer training is another inspiring trend. With the availability of user-friendly apps and affordable wearable tech, even amateur players can access personalized training programs. This broader reach fosters a greater appreciation for the science behind the sport, empowering more individuals to train smarter.

However, deploying AI for personalized training does come with challenges. One of the primary concerns is data privacy. With vast amounts of sensitive information being collected, the risk of data breaches looms large. Ensuring that all data is securely stored and ethically used is paramount.

Another challenge is ensuring the accuracy of AI-driven insights. Though highly advanced, AI systems are not infallible. Constant updates and human oversight are essential to maintaining the reliability of the recommendations generated by these systems.

Despite these hurdles, the potential benefits far outweigh the challenges. The fusion of AI and personalized training opens new avenues for player development, making training more efficient, effective, and engaging. It creates an environment where every player, regardless of their current skill level, can strive for continuous improvement.

As AI continues to evolve, it's conceivable that future innovations will further fine-tune personalized training programs. Biomechanics, dietary habits, even sleep patterns—every aspect of a player's life could soon be optimized for peak performance. What was once the realm of science fiction is now increasingly becoming an integral part of everyday training in soccer.

In summation, personalized training programs powered by AI are transforming soccer in ways we once thought unimaginable. They provide data-driven, customizable solutions that cater to the unique needs of each player, fostering holistic development. This cutting-edge approach has already started to redefine the standards of training excellence, promising a more dynamic, engaged, and effective pathway to soccer mastery.

Chapter 7:
AI-Powered Game Analysis

The realm of AI-powered game analysis is transforming the way soccer is studied and played, bringing a strategic depth previously unimaginable. With advanced video analysis tools, coaches and analysts can dissect every moment of a match, discovering patterns and tactical trends that were once obscured in the chaos of live play. This technology doesn't just stop at observation; it actively assists in crafting strategies tailored to exploit opponents' weaknesses, offering a dynamic and highly specific blueprint for victory. By leveraging AI's capability to process massive datasets with unprecedented speed and accuracy, teams can craft intricate game plans that adapt in real time based on the flow of the match. This isn't just about having more data; it's about transforming raw information into actionable insights, revolutionizing how decisions are made on and off the pitch. The integration of AI in game analysis is not just a leap forward—it's a fundamental shift in the landscape of soccer strategy, enabling a deeper understanding of the game and fostering an environment where innovation and tradition can thrive together. This chapter delves into the cutting-edge tools shaping this new era and the profound impact they're having on the beautiful game.

Advanced Video Analysis Tools

Advanced video analysis tools have surged to the forefront of AI-powered game analysis, revolutionizing how soccer matches are dissected and understood. These tools enable coaches, analysts, and other stakeholders to sift through vast amounts of visual data with unprecedented accuracy and speed. As the technological landscape matures, the capabilities and applications of these tools expand, offering deeper insights into game dynamics and facilitating informed decision-making.

In the past, video analysis in soccer was a labor-intensive task. Analysts had to manually go through hours of footage to identify key moments, a process prone to human error and oversight. Modern AI-driven tools have changed this narrative by automating many aspects of video analysis. Algorithms can now quickly scan match footage, identifying critical events such as goals, fouls, and offside situations, and even pinpointing intricate tactical maneuvers. This level of detail was previously unattainable and offers a layer of strategic intelligence that can make the difference between winning and losing.

The primary advantage of AI in video analysis lies in its ability to process large datasets swiftly. Machine learning models are trained on thousands of hours of footage, learning to recognize patterns and anomalies that might escape even the keenest human eye. For instance, these tools can analyze player positioning, ball movement, and team formations in real-time, providing immediate feedback to coaches during matches. This allows for tactical adjustments on the fly, a potential game-changer in high-stakes scenarios.

One of the key components of advanced video analysis tools is player tracking. By using computer vision techniques, these systems can monitor player movements, not just when they are on the ball, but throughout the entire game. This comprehensive tracking enables a detailed review of a player's spatial and temporal behavior on the field. Coaches can use this information to fine-tune training programs, focusing on areas where players can improve their positioning, stamina, or decision-making.

Beyond player tracking, these tools offer granular analysis of team strategies. By examining formations and tactical approaches, AI can provide insights into how well a team is executing its game plan. For example, if a team intends to use high pressing, video analysis can reveal whether players are maintaining their positions and closing down opponents effectively. Similarly, defensive setups can be scrutinized to ensure that players are maintaining their shape and covering critical areas of the pitch.

Moreover, AI-enhanced video tools can identify and analyze key moments in matches, creating highlight reels that are not just for entertainment but also for deep tactical study. These highlights can be tagged and sorted to emphasize various aspects of the game, such as set-piece effectiveness or transition play. By breaking down these moments, coaches can better understand the strengths and weaknesses of their team and their opponents.

The ability to analyze opponent strategies through video analysis tools is another invaluable asset. AI can break down footage of upcoming opponents, identifying their tactical preferences, key players, and frequent play patterns. This information can then be used to devise specific game plans designed to exploit the opponent's weaknesses. For instance, if the AI identifies a tendency for a rival team's defense to falter under high pressure, a coach can alter their approach to capitalize on this vulnerability.

One technology underpinning these tools is deep learning, a subset of machine learning that involves artificial neural networks. These networks can identify complex patterns in data and are particularly well-suited for image and video recognition tasks. By training deep learning models on extensive match footage, they can learn to detect subtle nuances in player behavior and team tactics, delivering insights that were previously out of reach.

Real-time analysis is yet another feature that sets advanced video analysis tools apart. During a match, these tools can provide instant feedback, helping coaching staff make informed decisions on player substitutions, tactical shifts, and more. For example, if a key player is showing signs of fatigue, as identified through movement analysis, the coach can be alerted to replace them, mitigating the risk of injury and maintaining optimal team performance.

The integration of these tools into everyday training routines has also transformed how teams prepare for matches. By leveraging video analysis, training sessions can be tailored to address specific weaknesses and reinforce the strengths identified through game footage. It allows for a more scientific approach to training, ensuring that every session contributes meaningfully to overall team performance.

One of the biggest benefits of AI-powered video analysis is its objectivity. While human analysts can be influenced by biases or emotions, AI looks at data devoid of subjectivity. This leads to more accurate assessments and more effective strategies. Teams can benefit from a clearer, unbiased perspective on their performance, helping to enhance player development and team cohesion.

However, the use of these advanced tools is not without its challenges. The accuracy of AI-driven video analysis depends heavily on the quality of the data it processes. Poor-quality footage or incorrect labeling during the training phase can lead to inaccuracies. Additionally, the sheer volume of data generated can be overwhelming, requiring robust data management systems to handle, retrieve, and interpret this information effectively.

The human element remains a critical component of video analysis. While AI tools provide the data and initial insights, the interpretation and application of these insights still rely heavily on the expertise and intuition of coaches and analysts. The best results are achieved when AI and human intelligence work in tandem, combining the objectivity and computational power of AI with the nuanced understanding and strategic vision of experienced professionals.

Looking ahead, the future of advanced video analysis tools in soccer is promising. As technology continues to evolve, we can expect even more sophisticated algorithms capable of providing deeper insights and predictive analytics. These advancements will undoubtedly push the boundaries of what is possible in soccer analysis, enabling teams to fine-tune their strategies with surgical precision and potentially revolutionizing the way the beautiful game is played.

In summary, advanced video analysis tools represent a pivotal development in AI-powered game analysis. By enabling detailed, objective, and real-time scrutiny of matches, these tools offer an unparalleled depth of insight. They provide coaches and analysts with the information they need to make data-driven decisions that enhance team performance, player development, and tactical execution. As technology continues to advance, the potential applications of these tools will only expand, cementing their place as indispensable assets in modern soccer.

Opponent Analysis and Strategy

In the fast-paced world of soccer, knowing your opponent can often be the key to victory. With the introduction of AI into game analysis, the process of understanding and developing a strategy against opponents has never been more precise or insightful. AI-powered tools have fundamentally transformed the way teams prepare for matches, offering coaches and players unparalleled access to data and insights.

Through advanced video analysis tools and machine learning algorithms, AI can break down hours of footage into digestible segments, pinpointing strengths and weaknesses of opposing teams with an accuracy that was previously unimaginable. This allows coaching staff to identify patterns and tendencies in an opponent's play. For example, a team might tend to attack more aggressively on the left flank or switch to a high press only during specific phases of the match.

With AI at their disposal, coaches can create highly specific game plans tailored to exploit the weaknesses of the opposition. Imagine knowing the percentage of successful dribbles by an opponent's star forward or the exact zones where a rival fullback is most likely to be caught out of position. By cross-referencing such detailed stats, AI equips teams with actionable intelligence that can be converted into effective tactics on the pitch.

Moreover, AI doesn't just stop at video; it incorporates a myriad of data sources including player biometrics, in-game statistics, and even social media sentiment. Collectively, these data points construct a comprehensive profile of the opponent, covering both technical skills and psychological states. For instance, AI could detect if an opposing goalkeeper performs poorly under crowd pressure, adding another layer to the strategic planning.

One of the most captivating aspects of AI in opponent analysis is real-time adaptability. In the past, mid-game adjustments were primarily based on the coach's instinct and experience. Now, with AI, teams can dynamically analyze what's happening on the field and adjust their game plan accordingly. Real-time AI-powered dashboards deliver insights on how well the current strategy is performing and suggest possible tweaks – whether that means tightening the defense to counter a sudden attacking surge or changing formation to create better passing lanes.

But it's not just the professional level that benefits from these technologies. Lower-tier teams and grassroots organizations now have access to tools that were once the exclusive domain of elite clubs. These AI systems democratize the technology, allowing more teams to partake in high-level tactical analysis. This supports a more competitive and engaging sport overall, as smaller teams can now punch above their weight with data-driven strategy.

Another transformative element is how AI predicts potential future scenarios, enabling teams to rehearse different match outcomes. By simulating various game conditions, AI can forecast how an opponent might respond to a specific style of play. Coaches can use these simulations to prepare their teams with rehearsed responses to different in-game situations, enhancing players' decision-making capabilities under pressure.

The feedback loop between AI's analytical prowess and on-field execution is also closing rapidly. Wearable tech and sensor data feed back into AI systems, offering immediate insights into how well the strategies are being executed. If a team's pressing strategy isn't producing the expected turnovers, AI can quickly suggest modifications, thus optimizing performance in real-time.

Furthermore, AI equips analysts with the ability to constantly refine and improve their tactical knowledge. As the AI systems learn over time, they become more adept at making nuanced observations that might elude the human eye. This continuous learning aspect enables teams to evolve their strategic planning season upon season, staying ahead of the curve.

The impact of AI on opponent analysis and strategy extends beyond just match days. In the preparation phase, AI can also analyze training sessions to improve player readiness against specific opponents. By simulating an opponent's tactics in practice, players become familiar with the kinds of challenges they will face, making them more resilient and adaptable during the actual game.

Additionally, AI systems can now integrate psychological and behavioral data, providing deeper insights into the mental state of opposing players. This can be particularly advantageous in high-stakes matches where mental toughness can make the difference. Knowing an opponent's psychological profile allows for strategic moves that exploit any detected vulnerabilities, such as pressing a player more vigorously who is likely to crack under pressure.

The advancements in AI-powered opponent analysis have also pushed the envelope in collaborative coaching. Different departments within a team—such as tactical, physical, and psychological coaching—now operate in a more integrated manner owing to the unified data insights provided by AI. This holistic approach ensures that every aspect of a player's preparation and performance is optimized, providing a comprehensive game strategy that encompasses physical readiness, tactical know-how, and mental fortitude.

Yet, all these technological advancements would be less impactful without the human element. While AI offers data, insights, and predictive models, it is the intuition and experience of the coaching staff that ultimately translates these numbers into effective action. The synergy between human intelligence and AI is critical, as it combines the best of both worlds to develop nuanced, adaptable, and effective strategies.

A fascinating future development lies in the potential for AI to develop entirely novel strategies that haven't been considered by human coaches. By analyzing vast amounts of historical data and current trends, AI might identify innovative playing styles that could revolutionize how the game is played. This presents an exciting frontier where human creativity and AI's analytical power converge, potentially giving birth to unprecedented tactical approaches.

In conclusion, the integration of AI in opponent analysis and strategy heralds a new era in soccer. AI's ability to analyze vast quantities of data, predict future scenarios, and provide real-time insights represents a quantum leap in tactical preparation and execution. From professional leagues to grassroots teams, AI democratizes access to high-level insights, enabling any team to refine and elevate their play. The future promises an ever-deepening collaboration between human expertise and AI, transforming opponent analysis and strategy into a sophisticated, finely tuned aspect of the beautiful game.

Chapter 8:
Fan Engagement and Experience

As soccer continues to evolve with AI's transformative capabilities, the experience for fans has reached new heights, blending advanced technology with personalized interaction. AI-driven chatbots and virtual fans are more than just convenient—they're essential for creating a seamless connection between clubs and their supporters, offering real-time updates, instant responses, and an engaging virtual presence. Personalized content, tailored specifically to individual preferences, ensures that fans feel directly involved and catered to, whether they're watching from a stadium seat or their living room couch. Interactive platforms foster a sense of community and camaraderie, allowing fans to discuss tactics, celebrate victories, and share their passion in more vivid and engaging ways than ever before. Ultimately, the fusion of AI and soccer isn't just enhancing the on-field game; it's revolutionizing the spectator experience, making it more dynamic, immersive, and satisfying for every fan.

Chatbots and Virtual Fans

The intersection of technology and soccer has led to a fascinating evolution in how fans interact with the sport and their favorite teams. As artificial intelligence permeates various facets of the game, it revolutionizes fan engagement and experience through chatbots and virtual fans. These technological marvels are reshaping the way fans participate, support, and connect with the game, providing personalized experiences that were once unimaginable.

Chatbots, often designed with AI-powered natural language processing, are increasingly becoming the face of customer service for soccer clubs. Whether it's answering frequently asked questions about match schedules, providing real-time updates, or guiding fans through the ticket purchasing process, chatbots are streamlining fan interactions. Imagine a fan inquiring about the latest team news; a chatbot can provide instant, accurate responses, saving time and enhancing the overall experience. The immediacy and availability of these virtual assistants ensure that fans remain engaged and informed 24/7.

But chatbots do more than just answer questions. They've evolved to provide personalized recommendations based on user preferences. For instance, if a fan frequently attends home games, the chatbot can notify them about ticket sales or special promotions tailored to their history. This level of personalization fosters a deeper connection between the club and its supporters, making each interaction feel unique and valued. It's not just about information; it's about creating a relationship that feels personal and exclusive.

Moreover, chatbots are now incorporating advanced features such as conversational AI to engage fans in more meaningful ways. Through these intelligent interactions, fans can participate in quizzes, polls, and even real-time discussions about ongoing matches. This dynamic engagement transforms passive viewers into active participants, bridging the physical distance and creating a virtual community united by their love for the game. The integration of gamification elements like rewards and leaderboards further incentivizes participation, making the experience both entertaining and rewarding.

The rise of virtual fans, particularly during the global pandemic, showcased how technology could replicate the stadium atmosphere digitally. With physical attendance restricted, clubs and broadcasters sought innovative ways to maintain the buzz and energy typical of game day. Enter virtual fans—AI-driven avatars that represent real supporters. These virtual representations were seen cheering, waving flags, and celebrating goals on screen, providing a semblance of normalcy in unprecedented times. It was a testament to how technology could bridge gaps, keeping the spirit of the game alive even in the absence of physical presence.

These virtual fans aren't just static images; they are often programmed to react in real-time based on the events of the match. For example, an AI might trigger a wave of celebration among virtual fans when a goal is scored or display collective disappointment in the event of a missed opportunity. This collective digital response amplifies the emotional highs and lows of the match, enhancing the viewing experience for those watching from home. It's a way of creating shared experiences even when fans are miles apart, fostering a sense of community and belonging.

Virtual fans also play a significant role in social media interactions, where AI-driven accounts simulate the excitement and anticipation typically seen among real fans. These accounts "comment" on match events, share memes, and interact with human fans, making the digital landscape more vibrant and engaging. Through these interactions, clubs can continue to drive conversations and keep the momentum going long after the final whistle. This seamless blending of the physical and digital worlds ensures that fan engagement never wanes, regardless of external circumstances.

Looking ahead, the potential applications of chatbots and virtual fans are vast and promising. Imagine integrating these technologies with augmented reality (AR) and virtual reality (VR) to create fully immersive match-day experiences. Fans could don VR headsets to find themselves virtually seated in their favorite stadium, surrounded by other avatars, and interact with fellow supporters in real-time. This blend of AI and immersive tech can replicate the social aspects of attending a match, making remote support feel much more local and immediate.

Further, AI-driven analytics could enhance these experiences by providing deeper insights into fan behavior and preferences. By analyzing interactions with chatbots and virtual fans, clubs can gain valuable data on what fans enjoy, what engages them the most, and what aspects of the experience could be improved. This data-driven approach can inform future strategies, ensuring that every fan feels heard and catered to in the ever-evolving landscape of fan engagement. Personalization will continue to be at the forefront, with AI tailoring experiences to meet the unique needs and preferences of each supporter.

On a broader scale, these technologies serve an essential role in global fan outreach. Soccer clubs have supporters scattered across various continents, and traditional methods of engagement could never reach such a dispersed audience effectively. Chatbots and virtual fans transcend geographical barriers, bringing fans from all around the world closer to the action. Whether it's providing real-time translations of social media interactions or offering unique content tailored to different regions, AI ensures that clubs can connect with their global fan base in meaningful ways.

As clubs continue to adapt and integrate AI-driven solutions, the focus remains on enhancing the fan experience and fostering a deeper connection between the game and its supporters. The goal isn't just to utilize technology for the sake of innovation but to ensure that every fan feels a part of the journey, no matter where they are. This commitment to inclusivity and engagement is a testament to the transformative power of artificial intelligence in soccer.

In closing, the emergence of chatbots and virtual fans heralds a new era of fan engagement and experience in soccer. These technologies offer unprecedented levels of interaction, personalization, and community building, ensuring that the passion for the beautiful game remains as vibrant and inclusive as ever. As AI continues to evolve, so too will the ways in which we experience and connect with soccer, promising a future where every fan, regardless of location, can be an integral part of the sport they love.

Personalized Content and Interactive Platforms

Artificial Intelligence (AI) is revolutionizing soccer not only on the field but also in the way fans engage with the sport. The emergence of personalized content and interactive platforms has fundamentally transformed fan experience, making it more immersive and engaging. These innovations address the individual preferences of fans, tailoring content to make each interaction unique and memorable. In this section, we explore how AI-powered personalization and interactivity are redefining fan engagement.

Imagine logging into your favorite soccer app and being greeted with highlights from your favorite team, upcoming match schedules, and even inside scoops about player preparations—all curated just for you. AI algorithms analyze your past interactions, browsing history, and even social media activity to offer tailored content. Whether you're a fan of Liverpool or Barcelona, the content you receive is designed to feel like a bespoke experience, reflecting your unique taste and interests.

Interactive platforms amplify this personalized experience by offering real-time engagement opportunities. Imagine watching a live match while participating in polls, quizzes, or even interacting with other fans. These platforms utilize AI to track live game data and deliver relevant statistics, enabling fans to engage with the game on a deeper level. This isn't just about watching a match; it's about being part of a community that lives and breathes soccer.

Fan engagement starts with content delivery. In the old days, you had to wait for the evening news or check the sports section of your newspaper to catch up on soccer updates. Now, your smartphone does all the work. With AI-driven algorithms, apps can curate content—from game highlights to tactical analysis—which can be delivered to your device in real-time. Personalized notifications ensure you never miss a critical moment, whether it's a last-minute goal or breaking news about player transfers.

Interactive features aren't limited to just content consumption; they extend to creating content too. Platforms like TikTok and YouTube leverage AI to suggest popular trends and editing tools, enabling fans to make content that resonates with the soccer community. User-generated content is then shared and amplified through AI algorithms, reaching wider audiences and creating a network of fan-driven media. This cycle of creation and consumption makes the fan experience dynamic and interactive, two qualities essential for maintaining long-term engagement.

AI and machine learning are also making their mark in virtual fan interactions. AI-powered chatbots are becoming increasingly sophisticated, capable of holding meaningful conversations with fans. These chatbots can provide instant updates, answer FAQs, and even offer recommendations based on user preferences. It's like having a personal soccer concierge available 24/7, enhancing user experience by providing immediate and accurate information.

The rise of virtual and augmented reality further enriches the fan experience. Imagine donning a VR headset and finding yourself in the middle of a packed stadium, watching a game live with fellow fans from around the world. AI helps to customize these environments, allowing users to choose different perspectives or focus on specific players. Augmented reality (AR) applications can overlay real-time stats, player information, or tactical formations onto your screen, turning your living room into an interactive soccer hub.

Beyond virtual experiences, AI is helping to foster real-world connections among fans. Social platforms integrated with AI can suggest local fan clubs or event meetups, allowing supporters to connect with like-minded individuals. By analyzing social behavior and preferences, these platforms can recommend gatherings in your vicinity, strengthening community ties and enhancing the collective fan experience.

Fan engagement isn't just about entertainment; it's also about specialized content that delves into the nuances of the game. Tactical analysis, for instance, is no longer confined to TV pundits. AI enables fans to access in-depth breakdowns of strategies and player performances, often in real-time during live matches. Advanced algorithms can process large datasets to highlight key moments, offering fans a more profound understanding of the game. This kind of specialized content is invaluable for a fan base that craves more than just surface-level engagement.

Another vital aspect of fan engagement is merchandise. AI analytics can predict trends in merchandise sales, enabling clubs to offer personalized product recommendations. Whether it's a jersey with your favorite player's name or exclusive limited-edition merchandise, AI helps to create a shopping experience tailored to individual preferences. This not only boosts sales but also strengthens the emotional connection fans have with their teams.

On match days, personalized content and interactive platforms reach their zenith. Ahead of the game, fans receive tailored previews, including team line-ups, player stats, and tactical insights. During the match, interactive features like live polls and real-time commentaries keep fans engaged. Post-match, fans receive personalized summaries, video highlights, and tactical breakdowns, ensuring that the experience doesn't end with the final whistle.

AI-driven platforms also cater to fantasy sports enthusiasts. Customizable fantasy leagues, player performance analytics, and predictive tools offer a highly personalized experience. Fans can track their fantasy team's progress in real-time and receive recommendations to optimize their line-ups. This level of engagement turns casual viewers into active participants, further deepening their connection to the sport.

Furthermore, AI introduces an element of gamification to fan engagement. Reward systems based on user activity, such as loyalty points for interacting with content or participating in polls, foster increased engagement. These rewards can be redeemed for exclusive content, merchandise, or even tickets to live games. Such initiatives transform passive fans into active contributors, constantly interacting with the platforms.

Personalized content and interactive platforms are reshaping what it means to be a soccer fan in the digital age. AI acts as the linchpin that brings various aspects of fan engagement together, from personalized content delivery and virtual experiences to interactive features and community-building initiatives. This transformation is creating a new paradigm in sports fandom where engagement is continuous, immersive, and deeply personalized.

As we look to the future, the potential for AI in enhancing fan engagement seems boundless. Emerging technologies will only deepen the personalization and interactivity available to fans, offering even more ways to connect with the game and their favorite teams. As we embrace these innovations, the line between the virtual and real worlds continues to blur, creating an entirely new dimension of soccer fandom that's richer, more diverse, and intensely engaging.

In conclusion, AI is breathing new life into fan engagement through personalized content and interactive platforms. By leveraging the power of AI, clubs and platforms can offer fans a bespoke experience that transcends traditional barriers. Whether through customized content, real-time engagement, or community-driven initiatives, AI is making soccer fandom more vibrant and inclusive than ever before. The digital revolution in soccer isn't just about technology; it's about creating a more connected and enriched experience for every fan, everywhere.

Chapter 9:
Enhancing Broadcasts with AI

In recent years, artificial intelligence has brought a transformative edge to soccer broadcasts, making them more dynamic and engaging for fans around the globe. Intelligent commentary systems are revolutionizing the way matches are narrated, providing real-time insights that go beyond human capability. These systems analyze millions of data points in seconds, offering nuanced perspectives and predictive analyses that captivate viewers. Additionally, AI-driven real-time statistics have become a staple in modern broadcasts, allowing fans to access detailed metrics and trends as the game unfolds. This integration of AI doesn't just enrich the viewing experience—it also provides a deeper understanding of the sport, satisfying both casual watchers and the most dedicated analysts. By enhancing every aspect of the broadcast, AI ensures that each match is not just seen but experienced on multiple levels, creating a richer, more immersive connection between the game and its audience.

Intelligent Commentary Systems

In the rapidly evolving landscape of sports broadcasting, the infusion of artificial intelligence (AI) is significantly transforming the way audiences experience soccer. A pioneer in this transformation is the implementation of intelligent commentary systems, which aim to augment traditional broadcasting by seamlessly integrating AI technologies. Such systems not only enhance the quality of commentary but also offer a multi-layered, deeply analytical view of the game that caters to varied audience segments.

Imagine watching your favorite team play, and instead of relying solely on human commentators who might miss nuances or make errors, you have an AI-powered system that detects, analyzes, and articulates each pivotal moment with near-perfect accuracy. This system evaluates player performances and game dynamics, providing insightful analysis that is both real-time and contextually relevant. For fans hungry for in-depth understanding, this is a revolutionary approach to experiencing the game.

At its core, an intelligent commentary system leverages machine learning algorithms and natural language processing (NLP) to generate and deliver commentary. These algorithms are trained on vast datasets containing historical match data, player statistics, and linguistic patterns used in sports commentary. The result is an AI that understands the game as much as it understands language, enabling it to provide detailed, accurate, and engaging commentary.

The impact of these systems is multifaceted. First, they democratize access to high-quality analysis. Traditionally, elite-level commentary and analysis have been restricted to the confines of high-profile broadcasters with substantial resources. AI changes this by making it possible for even smaller networks or online platforms to deliver top-notch commentary. This advancement can broaden access to insightful analysis, making it available to fans across different geographic and socioeconomic backgrounds.

Furthermore, intelligent commentary systems add a layer of personalization to the viewing experience. Viewers can choose the level of detail they prefer, whether they want basic play-by-play updates, tactical analysis, or even statistics about specific players during the match. Such systems can adapt to the viewer's preferences over time, using machine learning to refine and customize the delivery of information continuously.

Another significant advantage is the capability to maintain objectivity and reduce bias in the game commentary. Human commentators, despite their best efforts, can sometimes exhibit partiality, whether consciously or unconsciously. AI systems, on the other hand, deliver commentary based on data and programmed algorithms, minimizing personal biases and ensuring a more impartial representation of the game's events. This objectivity can be particularly crucial in closely contested matches, where unbiased commentary adds to the credibility of the broadcast.

The potentials of AI in commentary also extend to multilingual translations. Traditional commentary often limits non-native speakers, making it hard for global fans to engage effectively. Intelligent commentary systems can bridge this gap by providing accurate translations in multiple languages in real-time, thus broadening the global reach and making soccer more accessible and enjoyable for international audiences.

One of the marvels of intelligent commentary systems is their capability to seamlessly integrate with other AI technologies used in broadcasting. For instance, real-time statistics can be fed directly into these systems, allowing commentators to provide up-to-the-minute data on player performances, team formations, and other critical aspects of the game. This kind of dynamic interactivity was hitherto unimaginable and adds a new dimension to how fans understand and engage with the game.

Moreover, intelligent commentary systems can significantly reduce the workload of human commentators. In a typical broadcast setting, commentators are inundated with an overwhelming amount of data that must be processed and communicated to the audience in real-time. An AI assistant can handle the heavy lifting by sifting through this information and presenting key insights, thereby allowing human commentators to focus more on storytelling and adding color to the game's narrative.

Of course, as with any technological advancement, there are challenges and limitations to consider. One primary concern is the potential for over-reliance on AI, which could inadvertently reduce the human element that many fans cherish. The passion, anecdotal knowledge, and unique perspectives brought by seasoned commentators add a richness to the broadcast that may be difficult for AI systems to replicate fully. Integrating AI in a way that complements rather than overshadows human input will be key to its successful adoption.

Another challenge lies in the sheer complexity of natural language understanding and generation in the context of sports commentary. While AI systems are highly proficient in handling data and generating text, capturing the nuance, humor, and rhetorical flair of human commentators requires nuanced programming and extensive training. It necessitates an ongoing effort to refine these systems continuously, ensuring they not only inform but also entertain the audience.

Despite these challenges, the potential rewards of intelligent commentary systems are vast. Incorporating AI in this way not only enhances the viewing experience for fans but also sets the stage for further innovations in sports broadcasting. As AI technology continues to evolve, the capabilities of these systems will only expand, heralding a new era of intelligent, interactive, and immersive soccer commentary.

Looking ahead, the integration of AI in broadcasting could pave the way for even more sophisticated applications, such as virtual reality (VR) and augmented reality (AR)-enhanced commentary experiences. Imagine donning a VR headset and not only watching the game but also seeing overlaid analytical insights and commentary from an AI system, creating an experience that is both visually and intellectually enriching. The convergence of these technologies will likely redefine spectatorship, offering unprecedented levels of engagement and understanding.

In conclusion, intelligent commentary systems are revolutionizing how soccer is broadcasted. By offering real-time, data-driven insights, personalizing viewer experiences, and extending the reach of high-quality analysis, these systems are transforming the relationship between fans and the beautiful game. This innovation underscores a broader trend towards using AI to enhance fan engagement, making soccer not only more accessible but also more exciting and insightful than ever before.

Real-Time Statistics for Viewers

In the evolving landscape of soccer broadcasting, real-time statistics for viewers represent a revolutionary leap forward. No longer are audiences mere passive spectators; they are now active participants, armed with a wealth of data that provides a deeper understanding of the game. This transformation is largely driven by the integration of artificial intelligence (AI) into live broadcasts. Through sophisticated algorithms and machine learning models, broadcasters can deliver a continuous stream of statistics that enhance the viewing experience.

Imagine tuning into a soccer match and having immediate access to a player's speed, distance covered, and pass accuracy—all updated in real-time. AI systems collect and analyze these metrics instantaneously, making use of data captured from cameras, wearables, and other advanced sensors. Viewers can see how a forward's sprint velocity affects a play or how a midfielder's passing network helps break down defenses. This level of insight is not just fascinating for fans; it's transformative for understanding the intricacies of the match.

But how exactly does this technology work? At its core, AI-powered real-time statistics rely on a constant flow of data from the field. Multiple cameras and sensors track player movements, ball trajectories, and even environmental conditions like wind speed and field humidity. This raw data is fed into AI algorithms that interpret it, identify patterns, and convert it into meaningful statistics displayed on the screen. For instance, a heat map showing the areas of the pitch where a player has been most active can be generated within seconds, providing viewers with nuanced insights.

Furthermore, real-time statistics can be personalized for individual preferences. Viewers can choose to focus on specific players, teams, or types of data. Some might want to track a star striker's performance, while others may be interested in team formation shifts. With AI, broadcasters can offer customizable viewing experiences, allowing fans to tailor the information they see according to their interests. This personalized approach ensures that the broadcast is engaging for both casual viewers and hardcore analysts.

The impact of real-time statistics extends beyond just enhancing the viewing experience; it can also influence sports betting and fantasy leagues. Accurate, up-to-the-second data can give bettors and fantasy sports enthusiasts an edge, providing them with the information they need to make informed decisions. This dynamic infusion of real-time statistics adds another layer of excitement to the fandom, as people can react almost instantly to in-game developments.

One of the remarkable aspects of AI-driven real-time statistics is its educational potential. For younger fans or those new to the sport, having complex data broken down into easy-to-understand visuals can be incredibly enlightening. It demystifies the game, making its strategic complexities more accessible. A young fan watching a breakdown of passing efficiencies or defensive structures learns to appreciate the sport on a deeper level, fostering a new generation of informed and engaged supporters.

From a technical perspective, implementing AI-driven real-time statistics requires robust infrastructure and collaboration among tech firms, data providers, and broadcasters. The data's accuracy hinges on the quality of the sensors and cameras used, as well as the algorithms' precision in analyzing movements and actions. Significant advancements in computer vision and machine learning have made it possible to capture and interpret data at unprecedented speeds and accuracies. As technology continues to evolve, we can expect even more sophisticated and granular statistics.

Broadcasters have started incorporating augmented reality (AR) to highlight these statistics in engaging ways. Imagine seeing a player's running path visually overlayed on the screen during a live broadcast or an AR representation of ball possession metrics displayed in the corner. These enhancements make the data more digestible and visually appealing to viewers, further integrating statistics into the fabric of the broadcast.

Despite the immense benefits, there are challenges to overcome. Data overload can be a concern. Cramming too much information onto the screen can overwhelm viewers and detract from the enjoyment of the match. Therefore, balance is crucial. The key is providing insightful, relevant statistics without disrupting the flow of the game. AI can help here by intelligently selecting which stats to display based on the match's context and viewer preferences.

Ethical considerations also come into play. Sharing real-time statistics involves handling vast amounts of player data, raising questions about privacy and consent. It's imperative that all stakeholders—teams, players, and broadcasters—agree on data usage norms to ensure transparency and ethical data handling. Striking the right balance between innovation and privacy will be essential as the technology advances.

To illustrate the practical application, consider a high-stakes match where every sprint, pass, and tackle counts. AI algorithms can analyze the game in real-time and provide instant feedback to broadcasters on which key moments to highlight, accompanied by relevant statistics. Did a player's exceptional sprint lead to a game-changing goal? A real-time data overlay can show the build-up, context, and the physical effort involved, making the moment even more captivating.

Moreover, AI-driven real-time statistics are becoming a crucial tool for commentators. Equipped with deeper insights and contextual data, commentators can offer more nuanced analysis, adding depth to their commentary. When a viewer hears about a player's extraordinary work rate, having statistics such as distance covered and sprint speeds to back up that claim adds credibility and engagement. Experienced commentators utilize these statistics to educate while they entertain, creating a richer viewer experience.

The integration of real-time statistics is not confined to television broadcasts. Streaming platforms, mobile apps, and interactive platforms are also evolving to include these features. For instance, fans watching a match on a mobile device can access live statistics through an app, customize their viewing dashboard, and even participate in live polls and discussions based on the data being presented. This cross-platform availability is essential in today's digital age, where multiscreen engagement is the norm.

As the technology and methodologies behind real-time statistics become more refined, the future holds even more promise. Imagine predictive analytics showing the likelihood of an event occurring—like the probability of a team scoring in the next five minutes, based on current play patterns. These advancements could not only elevate the viewer's experience but also open the door to new forms of interactive content and fan engagement.

In conclusion, real-time statistics enabled by AI are revolutionizing the way soccer is broadcasted and consumed. This transformation provides fans with a richer, more engaging viewing experience, allowing for deeper insights into the game's nuances. From personalized data to educational content, the possibilities are endless. The key is to harness this technology responsibly, ensuring a balance between innovation and usability, while always keeping the thrill and beauty of the beautiful game at the forefront.

Chapter 10:
Case Studies in AI-Driven Soccer

Exploring real-world applications, this chapter delves into how top soccer clubs have harnessed the power of artificial intelligence to redefine success on and off the pitch. Showcasing a variety of case studies, we examine how elite teams like Manchester City, Bayern Munich, and Paris Saint-Germain have integrated AI into their strategic frameworks. From transforming player scouting methodologies to implementing advanced video analysis for opponent strategy, these clubs are setting benchmarks. By highlighting their journey, the triumphs, and the lessons learned, we offer an insightful look into the challenges they faced, such as data integration and acceptance among coaching staff. These stories serve as both inspiration and a blueprint for other clubs aiming to leverage AI's transformative potential in soccer.

Success Stories from Top Clubs

Artificial Intelligence is changing the landscape of soccer, providing unprecedented insights and advantages for clubs at every level. Among these, some of the world's top football clubs have emerged as trailblazers, their adoption of AI propelling them to new heights of success and redefining the sport's potential. Let's delve into these fascinating success stories, where innovation meets tradition, making the clubs not just participants but pioneers.

To begin, FC Barcelona, one of the most storied football clubs globally, has always been known for its innovative approach to the sport. In recent years, their commitment to leveraging technology has been evident. With the advent of AI, Barcelona took it a step further, working with IBM to implement the Watson AI platform. This tool provides them with real-time analytics, not just for game performance but also for fan engagement. By analyzing massive amounts of data, Barcelona can tailor their strategies down to the minute details, significantly improving on-field performance and staying connected with over 150 million fans worldwide.

Similarly, Liverpool FC implemented an AI-driven scouting and player performance system, revolutionizing their recruitment and training processes. Under the guidance of Michael Edwards, Liverpool's approach to data analytics evolved significantly. The club partnered with Statsbomb, an advanced data analytics company, to delve into metrics that traditional statistics couldn't capture. For example, using machine learning, Liverpool could identify potential players who excel in creating space—a metric not easily evident through conventional analysis. This led them to talents like Mohamed Salah, who has since been integral to their successes, including their 2018–19 UEFA Champions League triumph and 2019–20 Premier League title.

Moving across Europe, German juggernaut Bayern Munich has also embraced AI, particularly in player health and injury prevention. Partnering with Siemens Healthineers, Bayern Munich implemented an AI-powered system that predicts injury risks. Through cognitive data analysis and wearable sensors, AI assesses players' physical conditions continuously, providing crucial insights into whether they should train, rest, or undertake particular exercises to avoid injury. This system has significantly reduced the incidence of muscle injuries, ensuring that top performers like Robert Lewandowski stay fit throughout crucial parts of the season.

One cannot overlook Paris Saint-Germain (PSG), who have taken AI implementation to astonishing levels. PSG utilizes AI to guide nearly all facets of their operations. On the pitch, they use predictive analytics for tactical planning and player performance monitoring. Off the pitch, PSG has employed AI to understand fan behavior and engagement better. Through AI-generated insights, the club tailors marketing strategies, ticket sales, and social media interactions. They've also employed virtual reality and AI to enhance training sessions, allowing players to immerse themselves in simulations of potential match scenarios. This comprehensive integration of AI has solidified PSG's status as both a football powerhouse and an innovation leader.

Ajax Amsterdam, famed for its world-class youth academy, has also embraced AI to maintain its tradition of player development. Ajax uses AI to scout young talent and predict performance trajectories accurately. With the help of GPS trackers and complex algorithms, Ajax can evaluate how young players perform in various conditions, both during training and competitive matches. This approach has allowed Ajax to nurture talents like Frenkie de Jong and Matthijs de Ligt, ensuring their development paths are optimized for long-term success.

In Italy, Juventus has harnessed AI's power for performance analysis and fan experience. Their collaboration with Dream VR has led to the creation of virtual reality experiences that offer fans immersive viewing options for games. Furthermore, Juventus employs machine learning to analyze match footage, extracting insights that help in formulating superior training methodologies and game strategies. This tech-forward approach has not only kept Juventus competitive domestically but also helped expand their global fanbase.

Manchester City's City Football Group has extended AI's reach beyond just one team, applying it across multiple soccer franchises globally. Utilizing SAP's HANA cloud platform, they perform vast data analysis across all their teams, from New York City FC to Melbourne City. The AI systems allow shared learning and resource optimization across the group, ensuring that best practices in training, player management, and fan engagement are applied consistently. This broad application of AI has not only bolstered City's achievements on the field but also fostered a strong network of sister clubs aligned in their strategic missions.

On the far side of the Atlantic, MLS club Seattle Sounders FC have been employing AI extensively for fan engagement and matchday experiences. By integrating AI through their mobile app, the Sounders offer fans personalized content, predictive ticket pricing, and tailored matchday logistics. The seamless fan experience has not only enhanced home game attendance but also fortified the club's loyal fanbase.

From player development to match tactics and fan engagement, AI is clearly providing top clubs with a competitive edge that was previously unattainable. Each of these clubs showcases how the confluence of tradition and technological innovation can lead to sustained success. By understanding and utilizing AI's potential, these clubs aren't just advancing their own agendas; they're carving out a new era for the sport itself.

As AI continues to evolve, these success stories will likely yield even more groundbreaking developments. The experiences of these pioneering clubs offer inspiration and a blueprint for others to follow. By continuing to explore and embrace AI, the future of soccer looks brighter—and undeniably smarter. The beautiful game will only continue to evolve, enhanced by the relentless march of technology and innovation.

Lessons Learned and Challenges

In understanding how AI has reshaped the world of soccer, it's crucial to acknowledge both the lessons we've learned and the challenges that persist. AI's transformative power in soccer is undeniable, yet the journey has been fraught with hurdles and insightful revelations that offer valuable guidance for future endeavors.

One profound lesson is the importance of data quality. High-quality, well-structured data serves as the backbone for any AI application. Clubs and organizations quickly realized that poor data can lead to inaccurate models and misguided decisions. Collecting data is not enough; it must be curated, cleansed, and validated to ensure its reliability. This painstaking process has pushed organizations to invest heavily in infrastructure and talent capable of handling data with the required precision.

Moreover, the integration of AI into soccer has accentuated the need for collaboration between technologists and soccer professionals. Data scientists and AI experts cannot operate in silos; they need to work closely with coaches, analysts, and players. This interdisciplinary approach ensures that AI tools are not only technically sound but also practical and applicable in real-world scenarios. The mutual learning between these groups has driven innovation and increased the overall effectiveness of AI solutions.

However, collaborating across fields has not always been smooth. Language barriers and differing perspectives can stall progress. Technologists may struggle to understand the nuanced concerns of coaches, while soccer professionals may find it challenging to grasp complex algorithms. Bridging this gap has been an ongoing challenge, highlighting the need for specialized roles that can serve as liaisons between technical and soccer domains.

Another lesson learned is the critical role of ethical considerations in AI applications. Data privacy, algorithmic bias, and the potential to undermine sporting integrity have emerged as prominent concerns. Striking a balance between leveraging data for competitive advantage and maintaining fair play is a tricky, yet essential, task. Ethical frameworks and regulations have started to take shape, guiding how AI is employed in ways that respect individual privacy and uphold the spirit of the game.

The challenge of bias in AI models cannot be understated. Algorithms trained on biased data can perpetuate existing disparities, leading to unjust outcomes. For example, if scouting systems rely heavily on historical data, they might overlook emerging talent from less prominent regions or socio-economic backgrounds. Addressing this requires constant vigilance, iterative model improvements, and a commitment to inclusivity.

Real-time application of AI in live games has been another challenging frontier. The dynamic and unpredictable nature of soccer makes real-time data processing and decision-making particularly difficult. AI systems need to be robust and flexible enough to provide actionable insights without disrupting the flow of the game. This has led to innovative advancements in real-time analytics but also exposed the limitations of current technology in handling the complexity of live sports.

On the training ground, AI-driven insights have revolutionized player development. However, translating these insights into tangible improvements is not straightforward. Coaches must adapt to new methodologies and integrate AI recommendations into existing training routines. This adaptation process can be disruptive and requires buy-in from all stakeholders. Resistance to change, fear of technology, and the perceived threat to traditional coaching roles are challenges that must be managed delicately.

Injury prediction and prevention, while promising, has faced its own set of challenges. The human body is incredibly complex, and while AI can identify patterns indicative of potential injuries, the accuracy and reliability of these predictions are still evolving. False positives and negatives can lead to over-cautiousness or complacency, respectively. Continuous validation of these models is necessary to enhance their accuracy and utility.

AI's impact on fan engagement and experience has been transformative, yet not without challenges. Personalized content and interactive platforms have redefined how fans interact with the sport. However, ensuring these platforms enhance rather than diminish the fan experience is vital. Over-reliance on AI-driven interactions can lead to a loss of the personal touch fans cherish. Hence, finding the right balance between automation and human engagement remains an ongoing challenge.

Broadcasting has seen significant enhancements through AI, with real-time statistics and intelligent commentary systems enriching the viewing experience. Here, the challenge lies in delivering these advancements seamlessly. Technical glitches, data lags, and the sheer volume of information can overwhelm viewers and detract from their enjoyment. Simplifying the user interface and ensuring the technology is robust enough to handle high-stakes live broadcasts are areas needing constant attention.

As AI continues to evolve, the financial implications also present a double-edged sword. While AI can drive revenue growth through better player performance, enhanced fan engagement, and streamlined operations, the initial investment and ongoing maintenance of AI systems can be substantial. Clubs, especially those with limited resources, face the challenge of justifying these expenditures and seeing a tangible return on investment.

In recruitment and scouting, AI has introduced more objective metrics and predictive capabilities. Yet, it's clear that these tools are not foolproof. Human intuition and experience still play a critical role in assessing a player's potential and fit within a team's culture. AI can complement but not replace the nuanced understanding that seasoned scouts bring to the table. The challenge lies in merging these approaches effectively.

Player mental health is another area where AI has shown potential but also raised concerns. Monitoring psychological well-being through AI can provide early warnings of issues, allowing for timely intervention. However, this also raises questions about privacy and the potential misuse of sensitive data. Clear guidelines and ethical boundaries must be established to ensure that such technologies benefit players without compromising their autonomy.

The implementation of AI in soccer has also highlighted the need for continuous education and training. Both current and future professionals in soccer must be adept at using and understanding AI tools. This necessity has driven the inclusion of AI and data analytics in training programs for coaches, analysts, and players. However, the pace of technological advancement means that ongoing education is essential to keep up with the latest developments.

To sum up, the journey of integrating AI into soccer has been marked by significant achievements and noteworthy challenges. The lessons learned from these experiences emphasize the importance of data quality, ethical considerations, interdisciplinary collaboration, and continuous education. While challenges such as bias, real-time application, financial investment, and maintaining the human touch persist, they also drive innovation and improvement. As we move forward, these lessons will serve as a guide, ensuring that AI continues to enhance the beautiful game in meaningful and responsible ways.

Chapter 11:
Interviews with AI and Soccer Experts

In this chapter, we dive into captivating conversations with leading minds at the intersection of AI and soccer. From AI pioneers who are pushing the boundaries of what's possible in sports technology to seasoned coaches and analysts who have embraced these innovations on the field, the insights are both profound and practical. Experts discuss how AI tools are altering game strategies, improving player performance, and keeping fans more engaged than ever. These firsthand accounts provide invaluable perspectives, shedding light on both the promises and challenges of integrating AI in soccer. The revelations here don't just inspire curiosity—they lay down a roadmap for the future of the beautiful game, driven by intelligent technology.

Insights from AI Pioneers

In the evolving landscape of soccer, artificial intelligence (AI) has emerged as a game-changer, not only on the field but off it as well. The journey to integrate AI into soccer hasn't been an overnight endeavor. Instead, it has been driven by relentless innovators who saw immense potential in marrying technology with sports. These AI pioneers, through their groundbreaking work, have been instrumental in defining new paradigms and setting benchmarks that now guide how we understand and engage with the sport.

Among these trailblazers is Dr. Patrick Lucey, currently the Chief Scientist at Stats Perform. Dr. Lucey's contributions have been pivotal in AI's role in game performance enhancement and player analytics. His research focuses on leveraging machine learning algorithms to interpret vast amounts of match data, providing insights that were once unimaginable. What makes his work particularly fascinating is its applicability across various levels of soccer—from professional leagues to youth academies.

Lucey's vision is driven by a deep understanding of both the human and technical aspects of the game. He believes that AI should augment the decision-making capabilities of coaches and analysts rather than replace them. "It's about providing the right information at the right time," he often emphasizes. His commitment to ensuring the human element remains at the core of soccer analytics is a testament to how AI can be used responsibly and effectively.

Another notable figure is Professor Daniela Rus from MIT's Computer Science and Artificial Intelligence Laboratory (CSAIL). Her work intersects robotics and AI, offering unique insights into how these technologies can transform soccer training and development. One of her standout projects involved creating AI-driven robots that can simulate various playing styles, helping teams better prepare for upcoming matches. Rus's innovations are not merely about technology for technology's sake; they're about creating tangible impacts on how teams train and strategize.

The research carried out by Rus and her team demonstrates AI's capability to individualize training regimes based on players' unique strengths and weaknesses. Her philosophy is rooted in personalization, ensuring each player can achieve their best potential by receiving tailored training activities. This approach signifies a shift from traditional, one-size-fits-all methodologies to more nuanced, data-driven strategies.

Meanwhile, at OpenAI, researchers like Greg Brockman are pushing the frontiers of what AI can accomplish in soccer. Brockman co-founded OpenAI with the aim of ensuring that artificial general intelligence (AGI) benefits all of humanity. Within the realm of soccer, his work has been geared toward creating AI systems that can understand and replicate human-level playing capabilities. This involves intricate simulations and deep learning algorithms capable of deciphering complexities such as player movements, team dynamics, and even fan behaviors.

Brockman's vision expands beyond just game performance and player analytics. He is passionate about using AI to enhance fan engagement, creating interactive AI models that can offer personalized content and real-time insights during live matches. This innovation promises to revolutionize how fans consume soccer, making it a more enriched and engaging experience.

Robin Gross's work at FC Barcelona's Innovation Hub exemplifies the collaboration between academia, industry, and professional sports teams to harness AI's full potential. Gross has extensively studied how AI can optimize scouting and recruitment processes. Through machine learning algorithms and predictive metrics, her team has developed models that can identify burgeoning talent across the globe. Findings from her work have helped FC Barcelona maintain its competitive edge by recruiting promising young players who might have otherwise gone unnoticed.

Gross's methodologies involve sifting through massive datasets ranging from player statistics to video footage. The insights derived from these analyses assist scouts and coaches in making data-driven recruitment decisions, ultimately shaping the future composition of the team. For her, the ultimate goal is to integrate AI seamlessly into traditional scouting practices, creating a balanced approach that respects both human intuition and algorithmic precision.

The contributions from cybernetic experts like Tim Crawford, often dubbed the "Soccer Scientist," bring another dimension to AI's application in the sport. Crawford's research primarily focuses on wearable technology and its role in monitoring and enhancing player performance. His work is grounded in the belief that real-time data collected from wearable devices can be transformational. By understanding metrics such as player fatigue, heart rate, and movement efficiency, coaches can make informed decisions that optimize player performance and minimize injuries.

Crawford's innovations have found applications not just in professional leagues but also in grassroots soccer. He advocates for democratizing access to these advanced tools, allowing even amateur players to benefit from AI-driven insights. This widespread accessibility embodies his vision of making soccer a better, safer, and more inclusive sport.

Dr. Nandi Junaid, a specialist in AI-powered mental health support, represents a fusion of technology and psychology. Working at the intersection of AI and human well-being, Junaid's expertise is crucial in addressing the often-overlooked mental health aspects of professional athletes. Her systems use AI to monitor players' emotional and psychological states, offering timely interventions and support mechanisms. These tools help manage stress, anxiety, and other mental health issues that can affect performance.

Junaid's work underscores the holistic nature of soccer performance. By recognizing that mental health is as critical as physical fitness, her AI applications provide a more comprehensive approach to player well-being. This integrated method ensures that players are at their best, both mentally and physically, as they take to the field.

Lastly, we can't overlook the contributions from interdisciplinary researcher Dr. Roberta Cirillo, who has been at the forefront of integrating AI with big data for team management. Her comprehensive approach includes using AI to streamline team operations, from tactical planning to resource allocation. By analyzing a mix of structured and unstructured data, Cirillo's methods offer a 360-degree view of team performance and management.

One of Cirillo's significant advancements includes AI-driven dashboards that provide real-time updates and predictive analysis. This allows coaches and managers to make quick, informed decisions during games and training sessions. Her work has demonstrated that when AI harnesses the power of big data, the resulting insights can considerably enhance a team's strategic and operational efficiency.

These AI pioneers, through their diverse contributions, have laid a robust groundwork for the future of soccer. Their insights not only validate the transformative power of AI but also illustrate a pathway where technology complements the intrinsic human elements of the game. As AI continues to evolve, their pioneering efforts will undoubtedly remain pivotal in shaping the next chapters of soccer's ongoing narrative.

Perspectives from Coaches and Analysts

When it comes to integrating artificial intelligence into the realm of soccer, who better to provide insight than the coaches and analysts who live and breathe the sport? They offer a unique blend of tactical wisdom and firsthand experience with cutting-edge technology. This section dives into their thoughts, shedding light on how AI is reshaping the beautiful game from the perspective of those on the front lines.

Coaches, at their core, are the strategic masterminds behind successful teams. They're responsible for developing game plans, analyzing opponent strengths and weaknesses, and making real-time decisions during matches. The introduction of AI into their toolkit has been nothing short of revolutionary. "AI-driven analytics have given us new dimensions of understanding soccer," says one prominent coach. "It's like having another assistant coach who's impeccably accurate."

This sentiment is echoed by many others in the profession. AI's ability to process vast amounts of data in mere seconds allows coaches to make more informed decisions. For instance, AI algorithms can quickly identify a pattern in the opposing team's play or flag up fatigue levels in key players, enabling timely substitutions. Furthermore, AI doesn't just stop at analysis. It augments the decision-making process, providing simulation options for different tactical choices, enabling a more dynamic and responsive strategy.

From the analysts' standpoint, their world has drastically expanded. Traditional methods of player evaluation relied heavily on watching game footage and manually recording observations. Nowadays, AI facilitates advanced video analysis, breaking down every aspect of player performance—from running patterns to touches on the ball—with astounding precision. "The level of detail we can now attain is incredible," notes a leading soccer analyst. "It's not just about stats anymore; it's about insights that were previously unimaginable."

Indeed, insights have become deeper and richer. Analysts can now produce more nuanced reports that allow for better-targeted training programs and tactical plans. For example, if a player's shooting accuracy declines in the latter half of a match, AI can pinpoint the exact moments leading to that drop-off. This data can then be relayed to the coaching staff for immediate action or integrated into the long-term training regimen of the player.

Another fascinating area where AI has made considerable headway is in scouting and recruitment. Coaches and analysts together appreciate the way AI can sift through mountains of data to identify potential talent, sometimes even spotting prodigies that the human eye might overlook. "AI tools have been incredibly beneficial in discovering young talents in remote areas," observes a renowned youth academy coach. "They help us cast a wider net than ever before, broadening our reach without compromising on precision."

Despite these advancements, the human element remains irreplaceable. Coaches stress that while AI can provide data and suggestions, interpreting this data still requires a human touch. "At the end of the day, AI provides us the 'what' but we decide the 'why' and 'how'," says a veteran coach, elaborating on the collaborative nature of AI and human expertise. It's a symbiotic relationship where one enhances the other.

Moreover, the cultural shift towards embracing AI hasn't been uniform across all clubs and leagues. Some coaches are more open to adopting new technologies, while others remain cautious. An experienced analyst explains, "There's always a learning curve with new technology, and it varies from team to team. Some coaches adapt quickly, incorporating AI into their daily routine, while others take a more skeptical approach, integrating it slowly."

The blend of skepticism and enthusiasm is what ultimately pushes the boundaries of what's possible. Coaches and analysts often pilot new AI tools on a small scale before fully committing, using initial success as a springboard for broader implementation. When these pilots prove fruitful, the findings can ripple through the sporting community, providing valuable case studies that facilitate wider adoption of AI technologies.

Interestingly, the collaboration between humans and AI is fostering new skill sets among coaches and analysts. Data literacy is becoming increasingly important. Advanced analytical tools require understanding basic data science principles, and many professionals in the field are taking courses to get up to speed. "We're not just coaching; we're almost becoming part-time data scientists," says one forward-thinking coach, encapsulating the new duality in their roles.

In essence, AI has democratized information in soccer, leveling the playing field and offering smaller clubs resources that were once exclusive to elite teams. This leveling effect doesn't just alter game day strategy but extends into training, recruitment, and even fan engagement. The buzzword here is "accessibility"—and it's a game-changer.

On the flip side, seasoned analysts contend with challenges as well. There's always the risk of data overload or misinterpretation. In a sport as fluid and unpredictable as soccer, there's a fine line between actionable insights and background noise. As one analyst puts it, "Not all the data is useful. Knowing what to ignore is just as important as knowing what to focus on. Sometimes, the sheer volume of information can be overwhelming, potentially leading to analysis paralysis."

To mitigate these risks, many teams now employ dedicated data scientists to work alongside traditional analysts and coaches. This coalescence of skills ensures that data is not only collected but rigorously vetted and contextually understood. It's a holistic approach that maximizes the benefits while minimizing potential pitfalls.

Training programs have also seen substantial improvements due to AI-driven personalization. Each player has unique strengths and weaknesses, and AI helps tailor training regimens to suit individual needs. Coaches love the ability to monitor each player's progress in real-time, making adjustments as required. "AI's real-time feedback has significantly reduced our response time," says a club's head coach. "We can address issues the moment they arise, be it in training sessions or live games."

Ultimately, the conversation with coaches and analysts reveals a clear consensus: while AI is an extraordinarily powerful tool, it's the marriage between technology and human intuition that truly elevates soccer. The future is not about replacing human expertise but enhancing it. AI, with its unerring accuracy and capability to crunch vast amounts of data, provides a firm foundation. On this foundation, the creative and strategic elements of coaching and analysis can soar to new heights.

There's still much to explore and perfect in this AI-soaked era of soccer, but one thing is certain—the coaches and analysts who embrace this technological revolution while retaining their foundational skills will lead the charge. They're the navigators guiding their teams through uncharted waters, blending tradition with innovation, and ensuring that soccer remains as thrilling and unpredictable as ever.

Chapter 12:
Ethics and AI in Soccer

As AI continues to reshape the landscape of soccer, ethical considerations have swiftly emerged as critical points of debate. From the sanctity of player data to the potential for AI to tilt the playing field, there is an urgent need for clear ethical guidelines. For instance, data privacy looms large, as vast amounts of personal and performance information are collected and analyzed. Clubs and stakeholders must navigate these waters carefully to protect the rights of players while leveraging data for competitive advantage. Moreover, the broader implications of AI's influence—such as the essence of sportsmanship and the authenticity of human decision-making in the game—are sparking thought-provoking discussions. AI's potential to inadvertently introduce biases or overshadow human judgment necessitates a balanced approach, ensuring that technology augments rather than diminishes the heart of the beautiful game. By addressing these ethical concerns head-on, the soccer community can strive towards a future where AI enhances the sport without compromising its core values.

Data Privacy Concerns

As artificial intelligence (AI) continues to shape the landscape of soccer, an increasing amount of data is being generated, collected, and analyzed. This surge in data brings to the forefront pressing concerns about data privacy, especially in a field as competitive as professional sports. The deployment of wearable technology, the use of cameras for video analysis, and the proliferation of AI-driven tools have led to unprecedented levels of data about players, teams, and fans. While these innovations offer exciting opportunities, they also raise important ethical questions about how this data is managed, protected, and used.

One primary concern is the collection and storage of personal data, particularly that of players. Wearable devices can track a variety of metrics, including heart rate, movement patterns, and even psychological states. This data can provide invaluable insights for coaches and analysts to optimize performance and reduce injury risks. However, when sensitive personal information is recorded and stored, it becomes vulnerable to misuse or breaches. Who owns this data, and who has the right to access it? These questions are not merely theoretical; they have real-world implications.

Consider the scenario where a player's health metrics are leaked to competitors or the public. Such incidents can tarnish reputations, affect contract negotiations, and even lead to mental health issues. Therefore, it is imperative that soccer organizations implement stringent data protection measures, compliant with regulations such as the General Data Protection Regulation (GDPR) in the European Union. These regulations require organizations to obtain explicit consent from individuals before collecting their data and to ensure that data is stored securely and used responsibly.

Another aspect to consider is the transparency of data usage. Players, coaches, and fans alike deserve to know how their data is utilized. Are there clear policies that outline what kinds of data are collected, how they are used, and who has access? Transparency fosters trust, but it also ensures that individuals can make informed decisions about their privacy. For example, players might be more willing to share health data if they fully understand the benefits and risks involved.

The issue extends beyond just the data of players. Fans' data is also collected through various channels, including ticketing systems, fan engagement platforms, and social media interactions. Personalized content and interactive platforms can greatly enhance the fan experience, but they also require the collection of personal information such as location data, preferences, and even purchasing behavior. Here again, the need for transparent data policies and robust security measures is incredibly important.

Moreover, as AI systems become more integrated into scouting and recruitment, there are potential biases in data interpretation that can arise. AI algorithms learn from historical data, which may contain inherent biases. For example, if past scouting reports have a bias towards certain physical attributes or backgrounds, the AI could perpetuate this bias in future talent identification processes. Ethical AI deployment in soccer must include regular audits and updates to the algorithms to minimize such biases and promote fairness.

Given these complexities, soccer organizations must establish cross-disciplinary teams that include data scientists, ethicists, legal experts, and representatives from player associations. These teams can work together to develop comprehensive data governance frameworks that ensure the ethical collection, storage, and usage of data. These frameworks should also include protocols for data breaches, ensuring that there are clear and effective responses to any privacy violations.

Lastly, education plays a pivotal role in alleviating data privacy concerns. Players, staff, and fans should be informed about the kinds of data being collected and the measures in place to protect their privacy. This education can be achieved through workshops, informational sessions, and clear, accessible policies available on organizational websites. By fostering a culture of awareness and accountability, soccer can embrace AI innovation while respecting the privacy and rights of all stakeholders involved.

In conclusion, data privacy is not just a technical or regulatory issue; it is an ethical imperative that requires careful consideration and action. While AI offers tremendous benefits for performance optimization, fan engagement, and strategic decision-making in soccer, these advancements must be balanced with robust data privacy measures. Only then can we build a future for soccer that is both technologically advanced and ethically grounded.

The Debate Over AI's Role in Sports

As soccer continues to evolve, the incorporation of artificial intelligence (AI) has sparked significant debate among fans, coaches, analysts, and technologists. There's no question that AI brings a plethora of advantages to the game, but with it comes a host of compelling ethical concerns. This section delves into the crux of these debates, examining the benefits and drawbacks of AI in the context of soccer.

One of the primary arguments in favor of AI is its undeniable impact on enhancing game performance and player development. By leveraging AI technologies for real-time data analysis and tactical adjustments, teams can optimize their strategies during matches. This level of precision was unimaginable a few decades ago. Additionally, AI assists in scouting and recruitment through predictive performance metrics and automated talent identification. Such advantages are hard to ignore, making the case for AI incredibly strong.

However, the very aspects that make AI so appealing are often the same ones that raise ethical questions. Critics argue that the reliance on AI could overshadow the human elements that make soccer so captivating. The sport has always been celebrated for its unpredictability and the unique flair that individual players bring to the game. There's a concern that over-reliance on algorithms and data-driven decisions might strip away the soul of soccer, reducing it to a mere computational exercise.

Data privacy is another significant issue that fuels the debate over AI's role in sports. With the extensive use of wearable technology and AI-driven performance monitoring, a vast amount of personal data about players is collected. This includes not just physical stats but also deeply intimate metrics like mental well-being and psychological states. The potential misuse of such data raises considerable concerns about the invasion of privacy and the ethical responsibilities that come with data handling.

Additionally, the ethical implications of AI don't stop at privacy. There's ongoing worry about AI bias and its repercussions. For instance, algorithms developed based on historical data could inadvertently perpetuate existing biases in player selection and scouting. These biases could arise from various factors, including geographical regions, financial backgrounds, and even racial profile data. Addressing these biases is crucial to ensure fairness and inclusivity in the sport.

The democratization of AI tools might also exacerbate the existing inequalities in soccer. While top-tier clubs with substantial financial resources can afford state-of-the-art AI systems, lower-tier clubs and grassroots organizations may struggle to keep up. This digital divide could widen the gap between elite and amateur levels, creating an uneven playing field that contradicts the fundamental ethos of the sport.

Despite these challenges, proponents of AI argue that the technology can democratize the sport if implemented thoughtfully. Open-source AI tools and collaborations between tech companies and soccer organizations could help make advanced AI accessible to more teams. Moreover, AI's potential in youth development and grassroots soccer can also serve as a counterpoint to the argument about widening inequalities. The technology can provide young talents with opportunities to refine their skills, regardless of their socioeconomic status.

The ethical considerations extend beyond just players and teams. AI's influence on fan engagement is another contentious area. As AI-powered systems create personalized content and interactive platforms, they reshape how fans experience the game. While this enhances engagement, it also raises questions about data privacy and the commercialization of fan interactions. Balancing innovation with ethical responsibility is crucial to ensure that fans' data is handled with utmost care and respect.

At the organizational level, the integration of AI necessitates structural changes that demand careful ethical considerations. From training staff to adapting club policies, these transformations require a holistic approach that factors in the ethical implications of AI deployment. Clubs must navigate these changes while upholding the integrity and spirit of the sport.

One can't ignore the financial ramifications of AI in soccer. The technology presents opportunities for revenue generation, but it also necessitates substantial investments. The cost-benefit analysis must consider not just the financial ROI but also the ethical costs and benefits. This comprehensive evaluation is essential to make informed decisions that align with the broader goals of the sport.

In conclusion, the debate over AI's role in sports, particularly soccer, is multi-faceted and complex. On one hand, AI offers transformative benefits that can elevate the game to new heights. On the other, it poses ethical dilemmas that cannot be overlooked. Striking the right balance between leveraging AI for its advantages while addressing its ethical concerns is the key to ensuring that the sport remains true to its essence. As technology continues to advance, this debate will undoubtedly evolve, necessitating ongoing dialogue and thoughtful consideration from all stakeholders involved.

Chapter 13:
AI and Women's Soccer

A I is breaking new ground in women's soccer, revolutionizing every aspect of the sport from player development to fan engagement. By leveraging AI-driven insights, coaches can unlock new strategies and optimize performance in ways that were previously unimaginable. This technological advancement is not just leveling the playing field but setting new standards. Real-time data analysis and advanced video tools empower teams to fine-tune their tactics mid-game, while wearables and performance metrics facilitate customized training programs tailored to the unique physiology of female athletes. These transformative technologies are not confined to the elite tiers; they permeate grassroots levels, enabling young girls to benefit from the same advanced resources as their professional counterparts. As AI continues to penetrate women's leagues, it offers compelling case studies that demonstrate how barriers can be broken and new milestones reached, shifting perceptions and elevating the sport to unprecedented heights. The synergy between AI and women's soccer is a testament to how technology can champion inclusivity and drive progress in sports.

Breaking Barriers with Technology

Artificial intelligence (AI) is fundamentally reshaping many aspects of modern life, and women's soccer is no exception. Historically underutilized and overlooked, women's soccer has been gaining momentum worldwide, yet challenges persist regarding equity and opportunity. AI is playing a key role in breaking these barriers, providing tools and technologies that empower players, coaches, and fans alike.

One significant way AI is transforming women's soccer is through enhanced player performance monitoring. Wearable technologies, such as smart vests and GPS trackers, collect real-time data on players' movements, heart rates, and physical exertion. These devices feed information into AI-driven analytics platforms that offer detailed insights into each player's performance. These insights are invaluable for customizing training programs to fit each athlete's unique needs, helping them reach peak performance levels.

Imagine a coach strategizing based on real-time data from a match, adjusting tactics instantaneously to counteract the opponent's strengths and exploit their weaknesses. This type of real-time analysis was once reserved for men's leagues with extensive resources. Now, AI is democratizing access to these tools, bringing high-level strategic capabilities to women's soccer. As a result, teams can make data-driven decisions that were previously out of reach, leveling the playing field.

AI also brings advanced scouting and recruitment tools to the forefront, transforming how women's teams identify and nurture talent. Traditional scouting methods can be time-consuming and subjective. AI algorithms, on the other hand, can analyze vast amounts of data to pinpoint promising players, even in the most remote locations. Talent identification platforms can evaluate player performances from amateur leagues and youth academies, widening the talent pool for professional women's soccer teams.

Beyond the pitch, AI is enhancing fan engagement by creating more personalized and interactive experiences. Fan-centric technologies such as chatbots, AI-powered social media analysis, and personalized content recommendations ensure that supporters remain connected and engaged. This kind of engagement is crucial for growing the fanbase of women's soccer and increasing its visibility on the global stage.

Another groundbreaking application is in injury prediction and prevention. Women's soccer players are often at a higher risk for specific types of injuries, such as ACL tears. AI systems analyze data from wearables, previous injuries, and even genetic factors to predict susceptibility to injuries. With this information, tailored preventive programs can be created. Medical staff can adjust training loads, recommend specific exercises, and monitor recovery processes more effectively, extending players' careers and improving their quality of life.

The administrative realm also benefits from AI integration. Automated systems can assist in scheduling, logistics, and even marketing strategies. For instance, AI algorithms can identify optimal times for games to maximize attendance and viewership, or they can help craft targeted marketing campaigns to attract new fans.

Inclusivity and representation in sports media have always been critical issues. AI-driven video analysis and broadcast tools are enhancing the quality and reach of women's soccer coverage. These technologies provide detailed breakdowns of player performances and game tactics, which enriches the commentary and makes broadcasts more engaging. Enhanced broadcast quality, coupled with the rise of streaming services, ensures that women's soccer reaches a broader audience.

AI is also crucial in addressing unconscious biases that have historically plagued women's soccer. Advanced analytics platforms can provide objective performance metrics, reducing the influence of subjective opinions in player assessments. This ensures a fairer evaluation process based on data rather than preconceived notions, helping to dismantle long-standing biases in the sport.

The impact of AI extends to grassroots initiatives as well. Local leagues and community schools implement AI technologies to improve training and game strategies, making high-quality resources accessible to younger players. This not only builds a stronger pipeline for future talent but also fosters a community culture that values and supports women's soccer from a young age.

Ethics and data privacy are key considerations in the application of AI in women's soccer. Ensuring that data is collected, stored, and utilized responsibly is critical for maintaining trust among players and fans. Transparent data policies and adherence to ethical guidelines safeguard personal information, laying a strong foundation for AI technology's sustainable growth in the sport.

Collaboration between technology companies and soccer organizations is another area ripe for development. Partnerships that bring together AI innovators and women's soccer stakeholders can drive forward new solutions tailored specifically for women's teams. This collaborative synergy can accelerate the pace of technological advancements and their adoption in the sport, ensuring that women's soccer continues to evolve and thrive.

AI's role in breaking barriers is not limited to a single aspect but spans across the game of soccer, offering holistic advancements. From player health and performance to fan engagement and ethical considerations, the technology is a catalyst for change, aligning with the broader objectives of equity and inclusion in sports. The ongoing evolution of AI and its applications in women's soccer promises a future where opportunities are expanded and the sport can achieve its full potential, on par with men's leagues.

In summary, AI is not just a tool but a transformative force that is empowering women's soccer in unprecedented ways. It is driving performance, enhancing strategic decision-making, broadening talent pools, engaging fans, and promoting fairness. As these technologies continue to evolve, the potential to further break down barriers and transform women's soccer from the grassroots to the professional level is both immense and inspiring. The story of AI and women's soccer is one of innovation and progress, heralding a new era where technology and sport intersect to create a brighter and more equitable future.

Case Studies in Women's Leagues

Artificial intelligence is redefining the boundaries of women's soccer by harnessing data-driven insights and technological innovations. The ripple effect of AI in women's leagues isn't just noticeable; it's transformative. Teams and individual players are leveraging AI to refine their strategies, enhance performance, and engage fans at unprecedented levels. This section delves into various case studies where AI has made a considerable impact, helping women's soccer to gain deserving recognition and elevate its status globally.

Let's begin with an exemplary case from the National Women's Soccer League (NWSL). The Portland Thorns have utilized AI to streamline their scouting process. By integrating advanced machine learning algorithms, the Thorns can now assess and predict player performance with greater accuracy. Their AI systems analyze a multitude of metrics such as player speed, stamina, and decision-making capabilities, creating comprehensive profiles to aid in recruitment. This has enabled the Thorns to build a competitive roster, significantly enhancing their chances of winning championships.

Over in Europe, Olympique Lyonnais Féminin has become a front-runner in incorporating AI for injury prevention. Historically dominant both domestically and internationally, Lyon sought to maintain their edge through cutting-edge technology. By using wearable devices equipped with AI, they monitor players' physical conditions in real-time. These devices track biomechanics, heart rates, and even muscle fatigue, alerting medical staff to any anomalies. This preemptive approach has drastically reduced injury rates, allowing key players to remain fit throughout the season.

While the success on the field is significant, the Chicago Red Stars offer a compelling case for leveraging AI to enhance fan engagement. Utilizing AI-powered platforms, the Red Stars provide fans with personalized content. Chatbots manage real-time interactions during matches, delivering statistics, answering questions, and even predicting game outcomes based on current play. This innovation not only deepens the connection between the team and its supporters but has also attracted a broader audience, thus expanding the fan base.

Chelsea FC Women's team provides another insightful case study in AI-driven training regimens. The team uses AI to design individualized training programs for each player. Through machine learning, AI systems assess the players' strengths, weaknesses, and potential areas of improvement. These customized training plans continually adapt based on players' progress, optimizing skill development and physical conditioning. The result has been evident in Chelsea's performance, with players showing marked improvements in agility, precision, and resilience.

AI's influence isn't confined to elite clubs. Grassroots initiatives conducted by AI-focused organizations like StatSports have showcased how accessible technology can be. In partnership with various women's leagues around the world, they provide amateur teams with wearable devices that generate actionable insights. This democratization of technology means that even non-professional players can benefit from data-centric training insights, reducing the gap between grassroots and professional levels.

Barcelona Femení serves as a stellar example of incorporating AI in tactical analysis. The team employs AI systems to dissect gameplay footage, breaking down every move, pass, and defensive maneuver.

The AI offers real-time tactical adjustments, allowing coaches to make data-backed decisions on the fly. During their recent successful runs, AI facilitated instant replays and predictive analytics, giving coaches a detailed understanding of the opposition and optimal strategies to counteract them.

The Australian W-League's efforts to integrate AI have been nothing short of innovative. Sydney FC Women have adopted AI tools for talent identification. By analyzing local leagues and school-level competitions, they've identified rising stars before they become mainstream. These AI tools consider a blend of physical metrics, skill assessments, and even social factors like leadership qualities and team coordination skills. This forward-thinking approach ensures a steady influx of young talent, keeping the team competitive in the long run.

In the realm of mental wellness, the Paris Saint-Germain Féminine has taken significant strides. PSG uses AI to monitor the mental health of players, integrating psychological evaluations with team dynamics and individual performance metrics. AI systems detect early signs of stress or burnout, providing timely interventions through designated sports psychologists. This holistic approach helps to maintain a healthy, motivated squad capable of performing under pressure.

Japan's Nadeshiko League offers a unique case where AI has been instrumental in match scheduling and logistics. Leveraging machine learning models that predict weather conditions, travel constraints, and player wellness, the league is able to devise optimal match schedules. Such efficiency ensures minimal physical and mental stress for players, leading to more competitive and enjoyable matches. The fans, too, benefit from improved game-day experiences, fostering greater support for women's soccer in Japan.

When looking at officiating, the Frauen-Bundesliga in Germany showcases how AI can streamline referee decisions. AI-driven tools like automated offsides detection and foul recognition have been integrated into the VAR systems. The implementation has reduced human error significantly and enhanced the overall fairness of the game. This consistent and impartial officiating has elevated the integrity of the league, encouraging more fans to follow the sport with confidence.

Lastly, the application of AI in broadcasting has leveled the playing field for women's soccer. Networks that televise games from the Swedish Damallsvenskan use AI to generate real-time commentary and statistics. These intelligent systems enhance viewer experience by providing deeper insights and making the broadcasts more engaging. Such advancements are crucial for increasing viewership, consequently driving financial investments into women's leagues.

The transformative power of AI in women's soccer is multifaceted, from player development and injury prevention to fan engagement and tactical analysis. Each case study underscores AI's potential to not only enhance the sport but also ensure women's leagues receive the recognition they deserve. The pioneering efforts seen in these case studies reflect a promising future, where technology and sport converge to take women's soccer to unprecedented heights.

Chapter 14:
Youth Development and AI

The intersection of youth development and AI is transforming how we identify and nurture soccer talent from an early age. Traditional scouting, often reliant on the subjective eye of seasoned coaches, is now supplemented with sophisticated AI algorithms capable of analyzing vast amounts of data to pinpoint promising young players. These tools not only assess current performance but also predict future potential, allowing clubs to invest wisely in burgeoning talent. AI-enhanced youth academies are leveraging technology to create personalized training regimens, ensuring each player receives the optimal combination of drills, physical exercises, and strategic education tailored to their unique strengths and weaknesses. This bespoke approach fosters an environment where young athletes can maximize their potential, making the dream of becoming a professional soccer player more within reach for many. By integrating data-driven insights with traditional coaching wisdom, AI is revolutionizing the landscape of youth development in soccer, preparing a new generation of players to excel at the highest levels.

Identifying Talent Early

In the world of soccer, finding the next big star is akin to discovering a diamond in the rough. The sooner a promising talent is identified, the more time there is to nurture and develop the skills necessary to excel at the highest levels. Traditionally, talent identification relied heavily on the sharp eye and instinct of experienced scouts. However, artificial intelligence (AI) is revolutionizing this process, making it more scientific, comprehensive, and bias-free.

AI-driven tools are now capable of analyzing vast amounts of data from a young player's performance. This data comes from various sources like match footage, wearable tech, and even social media interactions. These tools leverage machine learning algorithms to discern patterns and insights that are often invisible to the human scout. For instance, subtle movements, decision-making speed, spatial awareness, and even physical growth potential can be quantified and evaluated.

Consider a youth tournament with hundreds of participants. Traditional scouting methods might cover a fraction of the players due to human limitations. AI, however, can examine every player's performance in detail. It can then rank these players based on predefined metrics, such as dribbling skills, passing accuracy, and defensive abilities. This level of scrutiny ensures that no potential star goes unnoticed.

Moreover, AI has the advantage of consistency. Human scouts can be subjective, influenced by personal biases or even momentary distractions. AI, on the other hand, applies the same criteria across the board, ensuring a fairer evaluation process. This democratization of talent identification means that players from underrepresented regions have a better chance of being discovered.

Another fascinating aspect of AI in talent identification is its predictive capabilities. Using historical data, AI can predict a young player's future career trajectory with surprising accuracy. Factors like injury history, psychological resilience, and adaptability to different playing styles are fed into the AI models. This allows clubs to make informed decisions on whether to invest in a young player.

However, it's important to recognize that AI is a tool, not a replacement for human judgment. While AI can identify skill sets and predict potential, the emotional and psychological elements of the game are harder to quantify. This is where experienced coaches and scouts still play a vital role. They bring in the human touch, understanding a player's mentality, passion, and drive, which are critical components for success in soccer.

Real-world examples are already demonstrating the effectiveness of AI in early talent identification. Clubs like FC Barcelona and Manchester City are at the forefront of integrating AI into their scouting systems. These clubs employ sophisticated AI models that analyze youth players across the globe, thereby unearthing hidden gems who might otherwise have slipped through the cracks.

AI isn't just for elite clubs either. Academies at all levels are beginning to adopt these technologies. Lower-league teams and even grassroots organizations are using AI to identify local talent, bringing more young players into professional environments. This broadens the talent pool, ensuring that soccer's future stars aren't limited to those who are fortunate enough to be born in soccer-centric regions.

The benefits of early talent identification extend beyond individual players and clubs. On a macro level, this leads to the elevation of the overall quality of the game. As more young talents are recognized and developed, the competitive standard across leagues and tournaments rises. This fosters a more exciting and dynamic sport, much to the delight of fans worldwide.

There are also societal implications. Young athletes from disadvantaged backgrounds often lack access to quality training and exposure. AI democratizes this by identifying potential irrespective of a player's socioeconomic status. As a result, soccer becomes more inclusive, offering opportunities based on merit rather than circumstance.

One compelling case is the use of AI by national teams during youth competitions. By incorporating AI-driven analytics, national federations can ensure they are nurturing the right talents from a young age. This holistic approach helps build more formidable national teams, capable of competing at the highest levels and achieving consistent success.

The AI systems used in talent identification are ever-evolving. As these systems become more sophisticated, they'll incorporate feedback loops and continually refine their algorithms. This means the accuracy and reliability of talent identification will only improve over time.

While the promise of AI in early talent identification is immense, it's also essential to approach its integration thoughtfully. Data privacy and the ethics of monitoring young athletes are paramount. Ensuring that all data collected is securely stored and used responsibly is crucial for maintaining trust in these technologies.

Moreover, clubs and academies must invest in training their staff to understand and effectively use AI tools. The insights generated by AI are only as valuable as the actions they inspire. Therefore, physiotherapists, coaches, and scouts must work hand-in-hand with data scientists to translate AI findings into practical development plans for young athletes.

To sum up, the advent of AI in identifying talent early is a game-changer in youth development. It brings precision, fairness, and accessibility to a process that was once nuanced and subjective. As AI continues to innovate and improve, its role in shaping the future stars of soccer will only become more pronounced, heralding a new era in the beautiful game.

AI-Enhanced Youth Academies

The integration of artificial intelligence into youth academies is transforming the way future soccer stars are nurtured. These AI-enhanced youth academies blend traditional coaching wisdom with cutting-edge technology, creating an environment ripe for developing young talent in more efficient and personalized ways than ever before. The impact is broad and multifaceted, revolutionizing how scouts identify promising players, how athletes train, and how coaches tailor their strategies.

AI-enhanced youth academies serve as breeding grounds for innovation. With AI-driven data analysis tools, coaches can now track and assess player performance metrics with unprecedented accuracy. These tools collect vast amounts of data from matches and training sessions, including positioning, speed, stamina, and tactical awareness. All of this data is then processed in real-time, allowing for immediate feedback that can guide training adjustments and specific drills to address any weaknesses.

One of the most compelling benefits of AI in youth academies is the emergence of personalized training programs. Traditionally, coaches have had to rely on a one-size-fits-all approach due to time and resource constraints. With AI, however, individualized training modules become feasible. Algorithms can analyze each player's data and craft training schedules that cater to their unique needs. For example, if a young midfielder struggles with ball control under pressure, the AI system would recommend specific drills to enhance this skill, ensuring the player gets the focused training they need to improve.

Besides personalized training, AI also plays a crucial role in injury prevention. Youth players often push their limits, sometimes leading to injuries that could have long-term impacts on their careers. Wearable tech equipped with AI capabilities can monitor stress levels and physical strain, predicting potential injuries before they become serious. Coaches and medical staff can intervene early, providing the necessary rest or modified training to prevent harm, ensuring that young athletes can train sustainably and stay on their path to success.

Scouting has also seen a paradigm shift with AI-enhanced youth academies. Traditionally, scouts would have to travel extensively to watch games and assess talent. Now, AI can process video footage from thousands of matches worldwide, analyzing key performance metrics and identifying promising talents based on predefined criteria. This automation allows scouts to focus on a more targeted list of prospects, ensuring that no stone is left unturned in the quest for the next soccer superstar.

Moreover, video analysis isn't just about picking the next star; it's also a learning tool for the current crop of academy players. Advanced video analysis tools powered by AI can break down footage of professional matches, highlighting strategic plays, player movements, and in-game decisions. Young players can study these sequences, learning from the best in the world and applying these insights to their own game. Through this method, AI helps bridge the gap between theoretical knowledge and practical application.

Communication between coaches, players, and parents has also been enhanced by AI technologies. AI-powered platforms offer a centralized hub for sharing performance data, training schedules, and development milestones. This increased transparency fosters a supportive environment, ensuring that everyone involved in a young player's journey is on the same page. Parents gain insights into their child's progress, while players can see quantifiable evidence of their development, which can be highly motivating.

Inspirational as it might be, the implementation of AI in youth academies isn't without its challenges. One of the primary concerns is ensuring data privacy and securing sensitive information. These academies must adopt stringent measures to protect personal data, often involving highly sophisticated encryption techniques and strict access controls. Ethical considerations also come into play regarding how much data is collected and how it gets used, especially when dealing with minors.

Another potential pitfall is the over-reliance on technology. While AI provides valuable insights and recommendations, it should supplement rather than replace human judgment. Coaches still play a vital role in interpreting data and making decisions based on a broader context, including emotional and psychological factors that AI may not fully understand. Striking the right balance between innovative technology and human intuition is crucial to the success of AI-enhanced youth academies.

Nevertheless, the potential benefits far outweigh these challenges. AI-enhanced youth academies can democratize access to high-quality training and development resources. Small clubs with limited budgets can use AI tools to level the playing field, ensuring that talented youngsters aren't overlooked due to a lack of resources. This democratization can foster a more diverse and inclusive environment where talent from varied backgrounds has a fair shot at reaching the professional stage.

The motivational aspect of AI-enhanced youth academies can't be overstated. Young players growing up in an AI-enriched environment are continually exposed to cutting-edge technologies, fostering a culture of innovation and relentless improvement. As they progress through the academy, they develop a mindset geared towards analytical thinking and adaptability—key attributes for succeeding in modern soccer.

Looking ahead, the integration of virtual reality (VR) and augmented reality (AR) into AI-enhanced youth academies seems inevitable. Imagine young players wearing VR headsets to practice their skills in simulated match environments, receiving real-time feedback from AI coaches during these sessions. AR can overlay tactical instructions onto the physical training field, helping players understand spatial dynamics and positioning in a way that traditional methods can't match.

In conclusion, AI-enhanced youth academies represent the future of soccer development. They provide the tools and methodology to nurture young talent more effectively, ensuring that promising players receive the best training, support, and opportunities to succeed. While challenges exist, the marriage of AI with traditional coaching has the potential to revolutionize how we develop the soccer stars of tomorrow. By embracing these technologies, we pave the way for a new generation of players who are not only skilled but also acutely aware of their own potential and capabilities.

Chapter 15:
The Role of Big Data

In the contemporary world of soccer, the role of big data has transformed from a mere supportive element to a cornerstone of decision-making processes. Integrating AI with big data allows teams to make more informed choices, whether it's about tactics, player management, or scouting new talent. The vast amount of data collected from various sources, such as match statistics, player biometrics, and even social media sentiment, is analyzed to gain deep insights that were previously unimaginable. This data-driven approach has ushered in a new era where decisions are made not on gut feeling, but on hard evidence, enabling teams to outperform their rivals in precision and efficiency. It's not just about collecting data; it's about understanding and leveraging it to reshape the game as we know it. From enhancing training sessions to creating personalized fan experiences, the applications are extensive and continually evolving, ensuring that the beautiful game remains ever more fascinating and competitive.

Integrating AI with Big Data

As the soccer world increasingly embraces technology, the integration of AI with big data has become a cornerstone in revolutionizing the sport. Both AI and big data independently hold significant value, but their combined potential opens up entire realms of possibility previously thought unachievable. With an intricate mesh of algorithms and vast datasets, AI can realize its fullest potential.

To put it simply, big data provides the vast pool of information while AI acts as the sophisticated tool that can analyze and interpret it. For years, teams gathered vast amounts of data: player statistics, game footage, GPS metrics, and much more. However, these data sets were often underutilized due to limitations in processing power and analytic techniques. AI changes this equation by introducing advanced methods like machine learning and neural networks to make sense of the data in real-time.

Imagine the wealth of tactical insights derived from analyzing over a million passes, tackles, and shots. AI algorithms can sort through these actions, identifying patterns and tendencies not merely as raw data points but as part of a larger tactical mosaic. For instance, the AI can reveal how often certain formations lead to successful goal-scoring opportunities or how player positioning impacts defensive solidity. This level of detail in analysis has become indispensable for many top clubs.

Take, for example, expected goals (xG) models, which have transformed how we understand shooting accuracy and goal likelihood. While traditional metrics might tell you a player has scored 10 goals in a season, an xG model takes it further by evaluating the quality of the chances they had. It considers the angle of the shot, the distance from goal, the type of pass received, and even the opponent's defensive positioning. Integrating AI with datasets that contain millions of shots enables teams to better evaluate both their attackers and defenders.

One of the fascinating aspects of integrating AI with big data is real-time decision-making during matches. Coaches and analysts don't have the luxury of conducting prolonged post-game analyses during halftime or even mid-match. AI, equipped with a rich database and powerful predictive models, can offer actionable insights in seconds. Algorithms can process live data feeds, compare them to historical benchmarks, and provide tactical recommendations to coaching staff—be it a player's substitution, a formation change, or a shift in tactical focus.

Moreover, player performance monitoring has received a massive upgrade through the marriage of AI and big data. Wearable technology captures every micro-movement on the pitch, feeding an endless stream of data into centralized systems. AI then analyzes this data to assess the player's fitness levels, workload, and even risk of injury. Over time, cumulative data can build a comprehensive profile for each player, helping club medical teams and coaches make informed decisions on player management. Personalized training programs, informed by an amalgamation of AI and wearables data, ensure that each player's specific needs and limitations are addressed, thereby enhancing performance and reducing injury risks.

Scouting and recruitment processes also benefit significantly. Automated talent identification systems, fueled by expansive databases and advanced computer vision algorithms, allow clubs to scout potential recruits from every corner of the globe. Instead of relying solely on the subjective opinions of scouts, clubs now have objective, quantifiable data to supplement their evaluations. AI-driven metrics can predict player potential, making the transfer market a less risky and more calculated exercise.

As if enhancing performance and scouting weren't enough, AI and big data integration also enriches fan engagement. Leveraging enormous quantities of fan data—from social media interactions to in-stadium behavior—AI-powered platforms can create personalized experiences that resonate on a deeper level with each fan. Whether it's customized content, interactive platforms, or even AI-driven chatbots offering instant responses, the harmony of AI and big data is steadily transforming how fans relate to the game they love.

Broadcasting is yet another area standing witness to this paradigm shift. Real-time statistics, insightful commentary, and immersive viewing experiences are all made possible by integrating AI into big data analytics. Fans watching a live match on television or streaming platforms can now see up-to-the-minute stats, informed predictions, and tactical analysis overlays that were unimaginable a decade ago. All this serves to make the game not just enjoyable but deeply engaging and informative.

Teams and entire leagues, armed with extensive data and AI-driven insights, are now capable of strategic decision-making at a level of granularity never seen before. The blend of AI and big data doesn't simply provide answers; it helps to ask better questions. Clubs now inquire not just about what happened or why it happened, but what could happen next and how they can best respond to it.

It's essential to recognize that the successful integration of AI and big data requires specific competencies and resources. Clubs need skilled data scientists, advanced computing infrastructure, and an organizational culture willing to embrace data-driven decision-making. Not every team can afford these resources, which creates a performance gap between the "haves" and the "have-nots." However, as technology becomes more accessible and costs decrease over time, the gap could narrow.

The role of big data in soccer isn't without challenges. Data privacy, particularly concerning player information, raises significant ethical questions. How do clubs balance the need for performance data with players' rights to privacy? As AI algorithms grow more sophisticated, they necessitate ever-larger datasets, escalating the stakes for proper data governance and ethical practices. Additionally, there is always the risk of over-reliance on AI. Human intuition, experience, and the unpredictable nature of sport remain irreplaceable elements. For AI and big data to be truly efficacious, they must complement rather than replace human decision-making.

Future prospects are undoubtedly exciting. The continued convergence of AI and big data promises even more transformative applications. Imagine predictive insights that can forecast trends across entire seasons, sudden tactical shifts to exploit an opponent's weakness in real-time, and even more personalized fan-engagement platforms that make every match an unforgettable experience.

In conclusion, the integration of AI with big data represents one of the most significant advances in modern soccer. It empowers teams with a level of intelligence that profoundly impacts performance, scouting, fan engagement, broadcasting, and much more. While challenges and ethical considerations must not be ignored, the collective potential of these technologies genuinely promises a bright and enthralling future for the beautiful game.

Applications in Team Management

Managing a soccer team is an intricate endeavor, one that involves aligning the unique skills and personalities of players with the overarching goals of the club. Enter Big Data, a transformative force that has led to groundbreaking changes in how teams are managed. By leveraging vast amounts of data, team managers can make informed decisions that were once guided by intuition alone.

For instance, player workload management is critical in preventing injuries and optimizing performance. Using data from wearables and tracking devices, managers can meticulously monitor each player's physical exertion. The data analysis can reveal patterns such as increased fatigue or higher risk of injury, enabling the coaching staff to make timely interventions. This proactive approach not only keeps players healthier but also ensures that they are in peak condition for crucial games.

Beyond physical management, data-driven insights extend to psychological well-being. Mental fatigue can be harder to quantify, but measuring variables like sleep patterns, social media activity, and even mood through wearable tech can paint a comprehensive picture. With AI algorithms processing this data, team managers can receive alerts about potential burnout or mental health concerns, facilitating timely support and interventions.

Big Data also plays a pivotal role in tactical management. During a game, advanced analytics tools can provide real-time data on both team and individual performances. Managers can fine-tune tactics based on objective measures such as possession statistics, passing accuracy, and spatial dynamics on the field. These insights allow for quick, data-backed decisions that can turn the tide of a match.

Furthermore, squad rotation, a critical aspect of team management, benefits immensely from data analytics. Balancing playing time to keep the squad fresh while maintaining peak team performance is no easy task. By analyzing various metrics including player form, fatigue levels, and opponent analysis, managers can optimize squad rotation strategies knowing they're backed by hard data rather than guesswork.

Another intriguing application is in scouting and recruitment. While traditionally dependent on scouts' observations and reports, the integration of Big Data allows for a more comprehensive evaluation of potential recruits. Algorithms can analyze thousands of data points from domestic and international leagues, providing a detailed profile of a player's strengths, weaknesses, and potential fit within the team's framework. This reduces the risk of transfer flops and ensures that new additions are strategically aligned with the team's long-term vision.

An often overlooked, but crucial, aspect is the optimization of training sessions. With detailed data on player performances and biometric readings, managers can customize training regimens to address specific weaknesses or build on particular strengths. For example, if data reveals a drop in a player's sprint speeds, focused conditioning drills can be incorporated to improve this aspect. Such tailored training ensures that every minute spent on the training ground is maximized for efficiency and effectiveness.

Communication within the team also gets a boost from Big Data. Clear, objective data can streamline discussions between coaching staff and players. Rather than relying solely on subjective observations, coaches can point to specific data points to guide their feedback. Players, in turn, benefit from concrete metrics that help them understand exactly what areas require improvement, fostering a transparent environment that facilitates growth and development.

Even administrative and logistical aspects of team management can be enhanced through Big Data. Travel schedules can be optimized to minimize fatigue, nutritional plans tailored to the needs of individual players, and even sleep schedules can be adjusted based on circadian rhythms tracked through wearables. These seemingly small adjustments can cumulatively make a significant impact on player well-being and performance.

Moreover, fan engagement and community relations also fall within the ambit of team management and are increasingly being influenced by data analytics. By understanding fan preferences through data analysis, teams can tailor their engagement strategies to build a loyal and engaged fanbase. This includes personalized marketing campaigns, interactive fan experiences, and even community initiatives that resonate with the fans' values and interests.

Big Data hasn't just changed the game on the field; it has revolutionized the business side as well. Contract negotiations, sponsorship deals, and merchandising strategies can all benefit from a data-driven approach. By understanding market trends and analyzing fan engagement metrics, team managers can negotiate better deals and craft strategies that maximize revenue while enhancing the brand's value.

The tactical boardroom is no longer a place for gut-feeling decisions. It's a data-informed environment where every strategy is backed by meticulous analysis. Whether it's tweaking formations, deciding on substitutions, or planning for the next transfer window, Big Data provides the precision and depth that modern soccer management demands.

In sum, the integration of Big Data into team management brings about a holistic approach that addresses both the tangible and intangible facets of managing a soccer team. From performance metrics and injury prevention to mental well-being and fan engagement, Big Data equips managers with the tools needed to navigate the complex landscape of modern soccer. The result is a more resilient, adaptive, and ultimately more successful team.

Chapter 16:
The Financial Impact of AI in Soccer

The integration of artificial intelligence in soccer isn't just transforming the game on the pitch; it's also reshaping the financial landscape of the sport. By optimizing player performance, refining scouting processes, and enhancing fan engagement, AI technologies drive cost efficiencies and open new revenue streams. Clubs can leverage predictive analytics to make smarter investment decisions, minimizing financial risks related to player acquisitions and training programs. Additionally, AI-powered platforms that personalize fan experiences attract lucrative sponsorship deals and increase merchandise sales. These financial benefits create a stronger, more sustainable economic model for clubs, ensuring they remain competitive in an increasingly data-driven industry.

Revenue Generation

As artificial intelligence (AI) continues to make inroads into various facets of soccer, its implications for revenue generation can't be overstated. From enhancing matchday experiences to creating new streams of income through data monetization, AI is reshaping how clubs and organizations think about profitability. At the core of this transformation lies the capability of AI to tap into vast amounts of data, decipher patterns, and offer actionable insights, which can then be leveraged to boost revenue.

One of the most immediate avenues through which AI is driving revenue is through fan engagement. AI-driven platforms can analyze fan behavior, preferences, and spending habits in real time. By understanding these nuances, clubs can tailor their marketing strategies to offer personalized experiences. This isn't just about personalized emails or notifications but extends to customized merchandise, ticket offers, and exclusive content. Such targeted marketing can result in higher conversion rates and enhanced fan loyalty, directly translating into increased revenue.

Broadcasting rights represent another lucrative revenue stream being revolutionized by AI. Advanced AI algorithms can generate real-time statistics and intelligent commentary, thus enhancing broadcast quality. When broadcasters incorporate AI-driven insights, they provide a richer, more engaging viewing experience. This enhanced product can command higher advertising rates and broadcast fees, significantly boosting revenue for both clubs and leagues.

Moreover, AI's predictive analytics capabilities enable clubs to optimize ticket pricing dynamically. By analyzing past ticket sales, weather conditions, opponent teams, and even social media chatter, AI models can predict demand for upcoming matches. Clubs can then adjust ticket prices in real time to maximize attendance and revenue. For instance, if a game against a lesser-known team on a rainy day is predicted to have low attendance, prices can be dropped or bundled with concessions to attract more fans.

AI is also opening up new revenue channels through the monetization of data. Clubs collect vast amounts of information, from player performance metrics to fan engagement stats. With AI, this data can be anonymized and sold to third-party firms interested in market research, advertising, and even sports betting. Such data partnerships provide a fresh revenue stream while ensuring privacy and compliance with regulations.

Sponsorship deals are yet another area where AI is making a considerable impact. Brands are always on the lookout for effective ways to reach their target audience, and AI offers the precision and depth of data necessary for highly targeted marketing campaigns. Clubs can use AI to provide detailed reports on how their fanbase interacts with different brands, allowing sponsors to tailor their campaigns accurately. This data-driven approach can help secure more lucrative sponsorship deals and foster long-term partnerships.

The in-game experience, enhanced by AI, also provides additional revenue opportunities. Smart stadiums equipped with AI can offer an array of services like facial recognition for fast entry, personalized food and drink offerings, and AR-driven experiences during breaks. These features not only improve the attendance experience but also open avenues for upselling and cross-selling. Imagine an AI system that recognizes a fan's drink preferences and offers a timely discount on their favorite beverage. It's small touches like these that can add up to significant revenue over a season.

Then there's the burgeoning field of eSports and virtual representations of soccer matches. AI can simulate games and tournaments, providing new content that attracts a different demographic. These simulations can be broadcast, sponsored, and even betted upon, creating a parallel revenue stream that operates independently of physical matches. For clubs with a strong brand presence, this can be an enormously profitable venture.

AI is also being utilized in the scouting and recruitment arena, making these processes more efficient and accurate. Clubs can save significant scouting costs by using AI to identify potential stars from around the globe. These cost savings can then be reallocated to other revenue-generating activities or simply add to the club's bottom line. It's a smart way of ensuring that every dollar spent provides the maximum return on investment.

Another fascinating development is AI's ability to optimize operational efficiency, leading to cost reductions that enhance profitability. For example, AI platforms can manage resources, schedule training sessions, and even predict maintenance needs in stadiums. By keeping operational costs low, clubs can ensure that a larger portion of their revenue is retained as profit.

AI also provides insights into fan sentiments through social media monitoring. By understanding what fans are talking about and what they care about, clubs can adjust their communication strategies and products to better align with fan interests. This kind of alignment often leads to increased merchandise sales and better fan attendance, both of which are fundamental revenue drivers.

It's not just about the immediate financial gains, either. AI offers long-term strategic advantages that contribute to sustained revenue growth. For example, AI can help in developing future leaders by identifying potential coaching talents and facilitating their growth. This ensures that the club remains competitive in the long run, which is crucial for maintaining and growing revenue streams.

AI is indeed transforming the soccer landscape in ways that were previously unimaginable. Whether it's through personalized fan experiences, enhanced broadcasting, dynamic ticket pricing, data monetization, or operational efficiencies, the potential for revenue generation is immense. Soccer clubs that embrace these technological advancements will not only stay ahead of the competition but also secure a more robust and diversified revenue base, ensuring their financial sustainability in the ever-evolving sports industry.

Cost-Benefit Analysis

As artificial intelligence (AI) continues to carve its niche in the world of soccer, evaluating the financial implications becomes not just important, but crucial. The cost-benefit analysis of AI in soccer offers a comprehensive way to understand what clubs, organizations, and stakeholders stand to gain or lose by adopting these cutting-edge technologies.

The initial investment in AI technologies can be significant. From acquiring advanced software to integrating sophisticated hardware like high-resolution cameras and wearables, the costs can quickly escalate. However, these upfront expenses need to be weighed against the myriad of long-term benefits AI promises. For instance, the use of AI in player performance monitoring systems can help prevent injuries, which in turn can save clubs millions of dollars in lost salaries and medical expenses.

When it comes to scouting and recruitment, AI offers a more refined and focused approach. Traditional methods often require extensive travel and manpower to scout potential talent. AI automates this process, reducing both the time and costs associated with talent identification. By using predictive performance metrics, clubs can make more informed decisions, ensuring that their investments in players yield high returns both on and off the pitch.

A key area where AI presents a strong cost-benefit ratio is in training and development. Virtual coaching assistants, for example, can deliver personalized training programs tailored to the unique needs of each player. This level of customization was previously unattainable and would have required a significant investment in personal coaching staff. With AI, the initial software costs are quickly offset by savings in manpower and the enhanced performance of athletes.

Moreover, AI-driven tools for game analysis and tactical adjustments bring additional savings and efficiency. These systems provide real-time data, allowing coaches to make informed decisions during the game. By leveraging advanced video analysis tools, teams can gain deeper insights into their opponents' strategies, thus improving their chances of winning. The competitive edge gained from these AI advancements translates directly into financial gain, whether through higher match attendance, merchandise sales, or sponsorship deals.

Fan engagement is another area where AI delivers a strong return on investment. Personalized content, interactive platforms, and AI-powered chatbots enhance the fan experience, making supporters feel more connected to their favorite teams. Increased fan engagement often leads to higher revenue streams from ticket sales, merchandise, and digital interactions. The relatively low cost of implementing these AI solutions, compared to the substantial financial returns they generate, makes this a worthwhile investment for soccer organizations.

The broadcasting sector also greatly benefits from AI integration. Intelligent commentary systems and real-time statistics enrich the viewing experience, attracting more viewers and higher advertising revenues. Although setting up these AI systems can be costly, the boost in viewer numbers and advertiser interest provides a substantial return on investment. Enhanced broadcasts create a more immersive experience, which in turn drives fan loyalty and generates additional revenue streams for clubs and broadcasters alike.

In summary, while the initial costs of implementing AI in soccer might seem daunting, the long-term financial benefits far outweigh these expenditures. The efficiencies gained, the enhanced performance of players, and the deeper fan engagement all contribute to a more robust financial outlook for clubs and organizations willing to embrace AI. The true value of AI in soccer lies not just in the technology itself, but in the far-reaching, transformative impacts it can have on every aspect of the beautiful game.

Chapter 17:
AI and Player Contract Negotiations

In a landscape traditionally dominated by gut instinct and extensive negotiations, AI is transforming how clubs and agents navigate player contract negotiations. By leveraging vast datasets and advanced analytics, today's clubs can more accurately determine a player's value based on performance metrics, potential growth, and market conditions. This shift not only streamlines the negotiation process but also minimizes the risk of overpaying or undervaluing talent. Agents, too, are adapting, using AI-generated insights to better advocate for their clients. What emerges is a more transparent and efficient framework, where decisions are backed by data, driving smarter investments and fostering equitable deals. The future of player contracts lies in balancing human intuition with the precision of AI, ensuring the beautiful game keeps evolving while preserving its core essence.

Data-Driven Valuation

In the realm of AI and player contract negotiations, the concept of data-driven valuation stands as a transformative force. Traditionally, the valuation of a soccer player hinged on a mélange of intuition, reputation, and subjective assessment. Scouts and managers relied heavily on their instincts, historical performance data, and future potential—a methodology that, while effective to a degree, was often riddled with biases and inefficiencies. Modern AI technologies have revolutionized this process, lending precision and objectivity to the intricate art of player valuation.

Data-driven valuation leverages vast datasets to craft a comprehensive profile of a player. These profiles include a range of metrics—covering physical attributes, in-game performance statistics, injury histories, and even psychological characteristics. Algorithms analyze these data points to predict future performance, offering a numerical valuation that accounts for a player's past achievements and future potential. This robust analytical framework empowers clubs to make more informed decisions when negotiating contracts, ensuring fairer deals for both clubs and players.

The backbone of data-driven valuation lies in machine learning models trained on historical data. By examining past transactions and outcomes, these models can recognize patterns in how certain attributes influence a player's market value. For example, a model might detect that players with exceptional dribbling skills and high work rates consistently secure higher transfer fees. Such insights enable clubs to identify undervalued talents whose contributions may not be fully captured by traditional scouting methods.

One of the most significant advantages of data-driven valuation is its ability to quantify elements of a player's game that are traditionally considered intangible. Aspects like leadership, situational awareness, and tactical intelligence are difficult to measure with standard statistics. However, advanced AI systems can analyze video footage, tracking player movements and decision-making processes to provide a deeper understanding of these qualities. This granular level of analysis helps clubs identify players who may excel in specific tactical systems or who possess the mental fortitude to perform under pressure.

Another key aspect of data-driven valuation is the integration of predictive performance metrics. AI models can forecast a player's future contributions based on various factors such as age, injury history, and playing style. For instance, a young player exhibiting rapid improvement might be projected to reach a higher peak performance, justifying a more substantial investment. Conversely, older players with recurring injuries might see their valuations adjusted to reflect the increased risk of future unavailability. This predictive capability enables clubs to tailor their investment strategies, balancing immediate needs with long-term potential.

Moreover, the use of AI in player valuations has illuminated the impact of positional value. Different positions on the field carry varying levels of importance and financial worth. AI can help quantify the contributions of each role, adjusting valuations accordingly. For example, a prolific striker who consistently scores goals might command a higher fee compared to a defensive midfielder who, while crucial to maintaining team balance, might not capture headlines. By understanding positional value through data-driven analysis, clubs can allocate their budgets more efficiently, investing in areas that will yield the greatest returns on both performance and financial fronts.

The influence of data-driven valuation extends beyond contract negotiations into broader financial strategies. Clubs can use AI insights to optimize squad depth and manage wage bills more effectively. Understanding the true value of players allows for more rational decisions regarding transfers, renewals, and releases. This strategic approach can result in a healthier financial footing for clubs, enabling them to compete more sustainably in the ever-competitive world of soccer.

Moreover, data-driven valuation has profound implications for the player transfer market. By standardizing valuations across different leagues and competitions, AI can facilitate more transparent and equitable transactions. Clubs from lower-profile leagues often struggle to receive fair compensation for their talent compared to their counterparts in more prominent competitions. Standardized, data-backed valuations can help bridge this gap, ensuring that players are valued fairly irrespective of their geographical or league context. This democratization of player value benefits the global soccer ecosystem, allowing talent to flow more freely and equitably.

Incorporating AI in player valuations also fosters greater accountability and transparency. The negotiations grounded in data can be scrutinized and justified with solid evidence, reducing the potential for disputes or accusations of favoritism. This transparency enhances the overall governance of soccer, promoting a culture of fairness and integrity. As clubs, agents, and players grow more accustomed to these data-driven negotiations, mutual trust can be strengthened, paving the way for more harmonious interactions within the soccer community.

However, it's not just clubs that benefit. Players, too, stand to gain from data-driven valuation systems. By understanding how they are assessed, players can make informed decisions about their careers. They can see areas where they need to improve to enhance their market value, making them more proactive in their development. This agency empowers players, allowing them to take control of their professional trajectories with a clearer understanding of how their performance translates into financial and career advancements.

In summary, data-driven valuation is revolutionizing the landscape of player contract negotiations. By harnessing the power of AI, soccer clubs can achieve unprecedented precision in valuing talent, leading to fairer and more strategic contract decisions. This transformation not only benefits clubs financially but also promotes a more equitable and transparent player market. For players, it offers a roadmap to career development grounded in objective, actionable data. As AI continues to evolve, its role in player valuation will undoubtedly grow, further harmonizing the intricate dance of soccer contracts with the rigor of data science.

Agent and Club Perspectives

In the electrifying world of soccer, player contract negotiations are among the most critical and intensely scrutinized aspects. Historically, these negotiations have been a mix of art and hard-nosed bargaining, driven by a player's performance on the pitch and market value. But the dawn of artificial intelligence (AI) has irrevocably altered this landscape. From an agent's perspective, AI provides a treasure trove of data that can be leveraged to advocate for players, while clubs see AI as a tool to make more informed, strategic decisions.

Agents, serving as player representatives, have always worn multiple hats: negotiator, marketer, and career advisor. With AI, their arsenal for player advocacy has expanded. They now have access to detailed performance metrics, historical data, and predictive analytics that can provide a more comprehensive valuation of a player. This kind of data presents a clear picture of a player's current form and future potential, making it easier to argue for higher salaries and better contract terms. No longer reliant solely on gut feeling or subjective opinions, agents can present clubs with data-driven evidence to support their demands.

For instance, consider an agent representing a young, emerging talent. Using AI-driven analytics, the agent can project the player's career trajectory by comparing them with historical data of similar players. By highlighting potential future performance, the agent can negotiate long-term contracts or lucrative transfer deals, bolstering their client's financial security and market standing.

On the other side of the negotiating table, clubs are equally excited about incorporating AI into contract discussions. Clubs have always sought to mitigate risks associated with high-value signings. AI tools offer an unprecedented ability to assess a player's health, performance consistency, and predicted career longevity. Real-time performance data and injury history analytics help clubs predict future injuries, thereby avoiding costly mistakes. These insights allow clubs to approach negotiations with a clearer strategy, often leading to more favorable contract terms for the club.

Moreover, clubs use AI to manage their wage structures more effectively. By aggregating data on player salaries across leagues, teams can benchmark what they are willing to pay individual players. They can offer competitive yet sustainable contracts, ensuring financial health while attracting top talent. This is particularly beneficial in today's soccer market, where transfer fees and wages have skyrocketed.

AI also aids clubs in determining the right moment to negotiate contract renewals. Predictive analytics can signal when a player's market value is at its peak, allowing clubs to either negotiate extensions or capitalize on transfer opportunities. Imagine a club facing the decision to extend a borderline star player or sell them. AI can assess the player's likely future performance and market dynamics, providing a well-rounded recommendation that factors in both athletic and financial considerations.

Furthermore, AI-driven sentiment analysis tools are starting to play a role in gauging public and fan sentiment around contract negotiations. For clubs that weigh fan opinion heavily in their decision-making, understanding how a potential new signing or contract extension might be received can be invaluable. This subtle layer of intelligence can make the difference between a smooth announcement and a public relations nightmare.

However, the integration of AI into player contract negotiations isn't without its challenges. One prominent concern is transparency. There is a delicate balance to be struck, as neither side wants to reveal too much of their analytics' inner workings. An agent might be wary of sharing all predictive analytics, lest the club identify potential long-term issues not previously disclosed. Similarly, clubs might not want to divulge the full extent of their data analysis to maintain a competitive edge.

Moreover, algorithmic biases can also affect negotiations. AI models are only as good as the data fed into them, which means historical biases in player evaluation and performance metrics can be perpetuated. For example, players from underrepresented leagues or those who defy traditional performance metrics may be undervalued by AI-driven models, making the agent's job tougher in securing the best deal.

Ethical considerations also emerge in AI-fueled negotiations. The amplification of performance metrics and health predictions in decision-making raises questions about the players' data privacy and mental health. As clubs lean more on technology, they must ensure this does not come at the expense of the player's well-being. Mental strain over analytics-based evaluations can lead to increased stress, affecting on-field performance and overall mental health. Agents and clubs need to develop strategies that consider these human elements to maintain a healthy, competitive environment.

Interestingly, the use of AI in negotiations also shifts some power dynamics. Agents who adeptly use AI are likely to find themselves in a stronger position, especially when representing players in the early stages of their career. Those resistant to adopting these technologies may find themselves at a disadvantage as clubs become more data-driven and less reliant on traditional scouting and evaluation methods.

The role of AI in negotiations also varies based on the club's resources. Top-tier clubs with substantial budgets for advanced analytics will naturally be more adept at leveraging AI in their negotiations. In contrast, smaller clubs may struggle to keep pace, resulting in a potentially uneven playing field. This could lead to a broader gap between wealthy and less affluent clubs, further intensifying the competitive dynamics of soccer.

Despite these challenges, the consensus among many agents and clubs is that the benefits of integrating AI into contract negotiations outweigh the drawbacks. As technology continues to evolve, so too will the methods of player valuation and contractual agreements. Both sides are learning to navigate this new landscape, developing best practices and uncovering new opportunities.

Ultimately, AI's influence on player contract negotiations promises to bring a new level of sophistication to the process. With precise data and predictive analytics, negotiations become less about posturing and more about intelligent, informed decision-making. Both parties stand to gain—agents through stronger cases for their clients and clubs through better-managed investments.

In summary, the perspectives of agents and clubs in the context of AI-driven player contract negotiations reveal a landscape rich with potential but also fraught with new challenges. Embracing this technology offers a path to more rational, data-backed decision-making while ensuring that the human elements at the heart of soccer are not lost in the process. As AI continues to evolve, so will the strategies and dynamics of player negotiations, shaping the future of the beautiful game in fascinating and unexpected ways.

Chapter 18:
Revolutionizing Refereeing with AI

Refereeing in soccer has always been fraught with controversy, but AI is set to change that narrative dramatically. By integrating advanced technologies like VAR (Video Assistant Referee), AI systems are already making significant strides in minimizing human error in officiating. These smart solutions offer real-time video analysis and instant feedback, making it possible to review contentious decisions with remarkable accuracy and speed. Beyond VAR, the potential extends to automated offside detection and foul recognition, bringing consistency and fairness to the game's officiating. As AI continues to evolve, it promises a future where the integrity of the sport is upheld with unprecedented precision and where debates over referee decisions become a relic of the past.

VAR and Beyond

Since its introduction, the Video Assistant Referee (VAR) system has brought a seismic shift to refereeing in soccer. This technology, designed to assist referees in making more accurate decisions, has its fair share of proponents and detractors. Nevertheless, its influence on the game is undeniable. VAR represents the cutting edge of AI in refereeing, serving as a precursor to even more advanced technological innovations that could further enhance the fairness and accuracy of soccer officiating.

One significant impact of VAR is the increased scrutiny and technological overlay on crucial game moments. Decisions on goals, penalties, and red cards, which often have game-changing consequences, are now subject to review. Yet, VAR is just scratching the surface. The scope of AI in refereeing goes far beyond video assistance, potentially ushering in a new era where human error is minimized through real-time data analysis.

Future iterations of AI in refereeing could include more sophisticated systems that leverage machine learning and computer vision. Imagine an AI capable of analyzing player movements and predicting fouls or offside positions with remarkable accuracy. Instead of relying solely on human judgment, which can be influenced by myriad factors, decisions could be corroborated by an AI that has parsed through countless hours of game footage and possesses an intrinsic understanding of the rules.

The notion of "intelligent whistles" is a tantalizing concept. Referees equipped with wearable technology that interacts directly with advanced AI systems can receive immediate feedback on decisions. For example, in instances of tight offside calls, the wearable could vibrate or signal the referee, providing a level of precision that's currently unachievable through human eyes alone.

Such innovations aren't merely theoretical. Various tech companies and research institutions are already developing complex algorithms capable of real-time movement analysis. These systems can instantly evaluate a player's position relative to the ball and other players to determine offside status or potential fouls, making the game fairer and more transparent.

Moreover, enhancements in natural language processing (NLP) can enable AI systems to understand the context behind players' and coaches' actions during high-pressure moments. An AI referee assistant could analyze the tone and content of interactions on the field, providing insights into potential unsporting behavior that may not be immediately apparent to a human referee.

This is not to say that human referees will become obsolete. Instead, AI serves to augment their capabilities, offering a second layer of decision-making that aids in maintaining the integrity of the game. Harnessing AI to reduce human error in refereeing is akin to fitting the game with a safety net, ensuring that crucial matches aren't marred by contentious decisions.

The journey beyond VAR also involves deep learning models that can predict and analyze player behavior based on historical data. Such models can flag tendencies and patterns, notifying referees of players who may be more prone to committing fouls or simulating them to gain an advantage. This preemptive approach could serve to deter such behavior, enhancing sportsmanship on the field.

There's also the exciting frontier of augmented reality (AR) and virtual reality (VR). Referees could be equipped with AR glasses that overlay vital information directly onto their field of vision. Real-time metrics, offside lines, and even instantaneous replays could be available at a glance, allowing for faster, more accurate decisions. Such innovations promise to make the referee's job easier and more consistent.

Technological advancements don't stop at officiating during the game. Off-the-pitch, AI can assist in the post-match analysis of refereeing performance. Machine learning algorithms can review game footage to identify mistakes and highlight areas for improvement. This feedback can be invaluable for training referees, ensuring that the standard of officiating continually rises.

Another aspect to contemplate is AI's role in addressing potential biases in refereeing. Unconscious bias—whether due to nationality, race, or team favoritism—can impact decision-making. AI, by its nature, offers an objective perspective, unclouded by personal biases. Though the AI itself requires stringent checks to avoid inheriting human biases, its use marks a step toward more impartial officiating.

However, the integration of AI in refereeing raises important ethical and logistical questions. Should AI systems have the final say, or merely assist human referees? How do we balance the immediacy and flow of the game with the pauses required for AI consultations? And importantly, how do we ensure that AI systems are transparent and accountable? These questions must be addressed as the technology evolves.

Public perception also plays a crucial role. Fans passionately debate VAR's merits and flaws, suggesting that the implementation of more advanced AI systems could face similar scrutiny. Thus, any technological enhancement must be accompanied by clear communication and education, ensuring that all stakeholders understand the benefits and limitations.

Looking ahead, the potential for AI in refereeing is boundless; its success depends on careful implementation, ongoing refinement, and an unwavering commitment to improving the game. As we evolve from VAR to more sophisticated AI tools, the ultimate goal remains the same: to preserve the essence and fairness of soccer, making it a better experience for everyone involved.

In conclusion, while VAR has already revolutionized refereeing, it serves as merely the first chapter in an evolving story. As AI technology continues to advance, we can anticipate a future where soccer officiating is more precise, fair, and consistent than ever before. This will require collaboration between technologists, governing bodies, and all stakeholders in the sport. The journey beyond VAR promises to not only enhance the integrity of the game but to preserve its passion and unpredictability—qualities that make soccer the beautiful game it is.

Reducing Human Error

Reducing human error in refereeing is essential for ensuring fairness, accuracy, and the overall quality of soccer games. Traditional refereeing, though rooted in years of experience and judgment, is inherently susceptible to inconsistencies. By integrating AI into refereeing, we can address several of these issues and bring about transformative changes.

A primary area where human error tends to occur is in making real-time decisions during fast-paced game play. Referees often have to make split-second calls that can drastically impact the outcome of a match. These decisions can be influenced by a range of factors, including the complexity of the action, the angle of view, and even psychological biases. AI, with its capacity for real-time data processing, can assist in making more accurate and consistent decisions.

Imagine an AI system that supports referees by analyzing every movement on the field with precision. Such a system uses multiple cameras and sensors to capture a holistic view of the game. Through machine learning algorithms, the AI can detect actions like fouls, offsides, and goal-line incidents almost instantaneously, providing referees with decisive information in a fraction of a second. This level of precision not only helps in making the right calls but also curtails the heated debates and controversies often associated with human errors.

Moreover, AI can manage retrospective data to identify patterns where errors are most likely to occur. For example, by analyzing hundreds of games, an AI system could highlight specific scenarios that frequently lead to incorrect decisions. This information can help referees prepare better and work on areas that require improvement. AI can also serve as an educational tool, providing tailored training sessions for referees based on their performance data, thereby continually enhancing their decision-making skills.

An example already in practice is the Video Assistant Referee (VAR) system, which utilizes video technology to review decisions made by the on-field referee. While VAR is a step in the right direction, it isn't without its flaws. Delays and interruptions in the flow of the game are common complaints. However, advancements in AI could make these reviews much more seamless and integrated, minimizing disruptions by quickly providing clear, data-backed decisions.

In addition to in-game applications, AI can reduce human error in post-match analyses and disciplinary actions. Automated systems can scan game footage for foul play that might have been missed during live play, ensuring that players are held accountable, even retrospectively. This could deter future misconduct, promoting a more disciplined and fair sporting environment.

Player safety is another critical area where AI can make a significant impact. Referees have the enormous responsibility of ensuring a safe playing environment, and missing a critical foul or dangerous play can have severe consequences. AI systems can swiftly analyze the impact and intent of player collisions, immediately alerting referees to any concern that needs attention. This not only reduces the risk of injury but also reinforces the importance of safe play among athletes.

Furthermore, integrating AI with biometric data collected from players can help in monitoring and predicting stress or fatigue levels. Fatigued players are more prone to committing fouls, often unintentionally. By understanding these metrics, referees can anticipate certain behaviors and make allowances, adjusting their oversight accordingly. This preemptive measure adds another layer of accuracy to refereeing, making the game fairer and safer.

One cannot dismiss the psychological impact of continuous errors on referees themselves. The pressure and scrutiny they face can lead to stress and burnout. AI can alleviate this pressure by sharing the decision-making burden and providing evidence-based support for their calls. This partnership between human intelligence and artificial intelligence fosters a more confident and relaxed mindset among referees, contributing to better overall performance.

Importantly, AI serves to enhance, not replace, the human element of refereeing. The goal is a symbiotic relationship where technology assists human judgment without undermining the referee's authority or instinctual understanding of the game. AI's role is to empower referees to make the best possible decisions for the benefit of the sport and its participants.

Reducing human error in refereeing through AI technology contributes to a more equitable and transparent soccer experience. By focusing on accuracy, consistency, and player safety, AI offers a transformative edge to one of the most pivotal aspects of the sport. As the technology continues to evolve, its ability to minimize human error will only get better, making the beautiful game even more beautiful.

Chapter 19:
AI and Mental Health Support for Players

The integration of AI in soccer isn't just about enhancing physical performance or strategizing for victories; it's also making significant strides in the realm of mental health support for players. Leveraging real-time data and machine learning algorithms, AI systems can now monitor a player's psychological well-being by analyzing behavioral patterns, social interactions, and biometric data. By identifying early warning signs of mental distress, these systems can prompt timely interventions, offering personalized support through AI-driven counseling tools and mental wellness apps. This proactive approach not only improves individual well-being but also fosters a healthier team environment, enabling players to perform at their peak both mentally and physically. As soccer continues to embrace technology, the focus on mental health shows a holistic commitment to nurturing athletes in every aspect of their lives, setting new benchmarks for what it means to be truly accomplished in the sport.

Monitoring Mental Well-being

In the dynamic world of soccer, players at all levels encounter numerous stressors, from rigorous training schedules to the high stakes of competitive matches. With the advent of artificial intelligence, monitoring mental well-being has emerged as an essential aspect of player support, promising to provide insights and interventions that enhance both performance and quality of life.

AI-driven monitoring systems have made remarkable strides in mental health assessment and management, primarily through the development of sophisticated algorithms capable of analyzing a plethora of data. These systems can collate information from wearable devices, psychological assessments, and even social media interactions. By interpreting this data, AI can identify patterns and anomalies that signify mental health issues, such as increased stress, anxiety, or exhaustion. This timely identification is crucial in preventing potential burnout and ensuring that players are mentally equipped to handle the pressures of their profession.

One of the significant advantages of using AI in monitoring mental well-being is its ability to process and analyze data in real-time. Wearable technology, such as heart rate monitors or smartwatches, captures physiological metrics that can indicate a player's stress levels. Coupled with machine learning algorithms, these devices can predict impending stress responses or emotional distress based on historical data patterns. This immediate feedback enables coaches and medical teams to implement interventions swiftly, ranging from rest periods to psychological support.

Furthermore, AI systems can benefit from natural language processing (NLP) to assess players' mental states. By analyzing languages, such as in player interviews or social media posts, NLP algorithms can detect signs of depression, anxiety, or other mental health issues. This linguistic analysis provides a non-intrusive way to monitor players, offering a layer of support that goes beyond traditional methods.

Another promising development is the integration of AI with virtual mental health assistants. These AI-driven platforms can offer personalized support and counseling to players, providing a confidential and accessible way for them to address mental health concerns. Virtual assistants can guide mindfulness exercises, cognitive-behavioral techniques, and even monitor progress over time. The personalized approach ensures that the support is relevant and effective for each player.

The benefits of AI in monitoring mental well-being extend beyond the individual player to the entire team. AI-generated insights can help coaches understand the overall mental climate of the team, allowing for more tailored training sessions and better team management. By knowing when the team is collectively feeling stressed or fatigued, coaches can adjust the intensity of training, plan team-building activities, or provide additional mental health resources.

Moreover, AI technologies have significant implications for injury prevention, which indirectly impacts mental well-being. Players who face injuries often experience mental health challenges due to the physical and emotional toll of recovery. By detecting signs of mental strain early, AI can help ensure that players are not only physically but also mentally prepared to return to the field, thus reducing the risk of re-injury.

Despite its advantages, AI-driven mental health monitoring is not without its challenges. One of the primary concerns is ensuring the privacy and security of the collected data. The intimate nature of mental health information requires stringent measures to protect it from unauthorized access and misuse. Establishing clear guidelines and ethical standards for data handling is essential to maintaining players' trust and ensuring the responsible use of AI technologies.

Another challenge lies in the accuracy and reliability of AI interpretations. Mental health is complex and multifaceted, and not all nuances can be captured through data alone. Therefore, AI systems should complement, not replace, traditional mental health support mechanisms. Collaboration between AI experts, mental health professionals, and sports scientists is crucial to developing holistic approaches that cater to the diverse needs of players.

As we look to the future, the integration of AI in monitoring mental well-being offers a promising avenue for enhancing player support. Continuous advancements in machine learning, data analysis, and wearable technology are likely to yield even more sophisticated tools for mental health monitoring. These innovations will play a pivotal role in creating a more supportive and balanced environment for players, ultimately contributing to their overall performance and success.

In conclusion, AI stands at the forefront of revolutionizing mental health support in soccer. By harnessing the power of data and advanced analytics, AI provides timely and personalized insights that are essential for maintaining players' mental well-being. While challenges remain, the potential benefits of AI-driven monitoring systems are immense, promising a future where players can perform at their best, both mentally and physically.

AI-Driven Psychological Support

As the demands on professional soccer players continue to intensify, the importance of mental well-being has come to the forefront. The grueling schedules, constant pressure to perform, and the need to maintain peak physical and mental conditions are tremendous. Here, AI steps in as an indispensable ally, offering psychological support that's both tailored and timely.

One of the most profound advancements in AI-driven psychological support is the ability to provide personalized mental health resources. Through sophisticated algorithms, AI can analyze vast amounts of data, including behavioral patterns, social interactions, and even biometric signals. These analyses generate insights into a player's mental state, allowing for interventions that are custom-fit to their individual needs.

Consider a player going through a rough patch following an injury. Decoding the subtle signs of depression or anxiety early on can make a massive difference. AI can help monitor these signs by analyzing variances in biometric data collected through wearables. Heart rates, sleep patterns, and even variations in speech can signal psychological distress. When irregularities are detected, AI systems can alert the medical staff and recommend steps for early intervention, thus helping to mitigate the problem before it escalates.

Moreover, AI-powered chatbots are revolutionizing how players receive psychological support. These chatbots offer a 24/7 lifeline, providing immediate assistance through conversational interfaces. They can conduct initial screenings, provide coping strategies, and even refer players to human psychologists for further help. The ability to have a judgment-free, always-available companion can be incredibly comforting, especially during moments of isolation or stress.

It's not just about real-time interventions, either. Longitudinal studies enabled by AI can track a player's mental state over months or even years. By doing so, patterns and triggers of psychological issues can be identified, offering deeper insights that are often missed in traditional mental health assessments. For instance, the data might reveal that a particular player experiences increased stress levels leading up to significant games or immediately following them. Armed with this knowledge, coaches and support staff can develop customized mental resilience programs to help the player better navigate these critical periods.

Additionally, AI has proven invaluable in creating simulated environments where players can mentally prepare for high-pressure scenarios. Through virtual reality (VR) experiences augmented with AI, players can practice handling stress-inducing situations in a controlled, safe environment. Whether it's preparing for the hostility of an away game or the pressure of a penalty shootout, these simulations can fortify a player's mental resilience, better equipping them for real-world challenges.

The inclusion of AI in psychological support also extends to enhancing team dynamics. By analyzing interpersonal relationships within the squad, AI can highlight potential conflicts or areas for improvement in communication. These insights enable coaches to address underlying issues proactively, fostering a more harmonious and supportive team environment. For example, if data reveals a lack of synergy between certain players, targeted team-building exercises can be employed to strengthen those bonds.

Furthermore, AI-driven sentiment analysis can help understand the public and media's impact on a player's mental health. By scanning social media, news articles, and public commentary, AI can gauge the tone and sentiment directed at individual players. This information can be critical for mental health professionals in tailoring their support strategies. Positive reinforcement can be emphasized, while strategies to cope with negative feedback can be reinforced, all tailored to the individual needs of the player.

It's also worth noting the scalability of AI solutions in mental health support. Traditional psychological services often face limitations in terms of availability and accessibility. However, AI can bridge this gap by providing consistent, high-quality support to players at all levels, from youth academies to professional leagues. The democratization of mental health resources ensures that even those at the grassroots level benefit from the same advanced support systems as their professional counterparts.

Imagine a young player in an academy showing early signs of burnout. AI can detect these signs through a combination of performance data, social interaction metrics, and self-reported feedback. Immediate and personalized interventions can be crafted, addressing both the psychological and physical aspects of their well-being. By catching these issues early, AI helps cultivate a healthier, more resilient next generation of players.

In conclusion, AI-driven psychological support marks a monumental shift in how mental health is approached within the realm of soccer. It goes beyond mere data analysis, offering holistic, personalized, and proactive support systems. The technology not only aids in detecting and managing mental health issues but also in building mental resilience and fostering a supportive team culture. As we continue to integrate AI into the beautiful game, its potential to enhance players' mental well-being and overall performance stands as a testament to its transformative power.

When integrated effectively, AI-driven psychological support can contribute to longer, healthier careers for players, improving their quality of life both on and off the field. It's more than just a technological advancement; it's a paradigm shift towards a more compassionate and comprehensive approach to mental health in soccer.

Chapter 20:
AI in Grassroots Soccer

The impact of artificial intelligence on grassroots soccer is transformative, ushering in a new era for amateur leagues and community initiatives. From local clubs to weekend teams, AI is leveling the playing field by providing access to sophisticated data analytics and training tools that were once the preserve of elite clubs. For instance, AI-driven apps help coaches design personalized training programs, monitor player performance, and make strategic adjustments in real-time. This democratization of technology enhances player development, fosters community engagement, and elevates the overall quality of the game. By integrating AI into grassroots soccer, we're not just equipping the stars of tomorrow but also enriching the experiences of players and fans at every level.

Transforming Amateur Leagues

For decades, amateur leagues have been the backbone of local communities, nurturing talent and fostering a love for soccer at the grassroots level. Traditionally, these leagues operated with limited resources, often relying on volunteer coaches and minimal technological support. But the landscape is changing rapidly with the advent of artificial intelligence (AI), which is bringing tools and capabilities once reserved for professional teams to the amateur scene. The infusion of AI into these leagues is not just revolutionizing the

Kick & Code

way the game is played and coached; it's transforming the entire ecosystem.

At the heart of this transformation are AI-driven technologies that empower amateur teams to analyze performance with unprecedented depth. Consider real-time data analytics, which tracks player movements, ball possession, and various other metrics during matches. This immediate feedback allows coaches and players to understand what's happening on the field in granular detail. No longer restricted to manual observation, they can adjust tactics on the fly, honing in on what works and what doesn't, right during the game.

One of the most impactful applications of AI in amateur leagues is in scouting and recruitment. Historically, discovering talent at the amateur level was a matter of chance—a talented player might catch a scout's eye during a local match. Now, AI systems scan through massive databases of player metrics, identifying emerging talents based on performance statistics and predictive models. These systems can identify attributes that might go unnoticed by the human eye, particularly in young players who are still developing their skills.

Imagine a small-town soccer club using AI to analyze video footage of their matches. Advanced video analysis tools can break down each play, providing insights into individual and team performance. Elements such as player positioning, passing accuracy, and defensive setups are meticulously parsed. This kind of analysis, which might have taken hours of manual review by coaches, is now available almost instantly. Moreover, cloud-based platforms allow these insights to be shared easily with players, who can access detailed feedback and tailored training plans directly from their devices.

Artificial intelligence is also revolutionizing training methodologies. Virtual coaching assistants, for instance, are becoming accessible to amateur teams, offering personalized drills and practice routines. These AI-powered tools assess each player's strengths and

155

weaknesses, recommending exercises that target specific areas for improvement. Athletes receive real-time feedback through wearable devices, continually fine-tuning their techniques based on data-driven recommendations.

Wearable technology is another game-changer at the grassroots level. Devices that monitor heart rate, physical exertion, and biomechanical movements provide valuable data that informs training regimens. Coaches can ensure players are training at optimal levels, preventing overuse injuries and improving overall fitness. These wearables, coupled with AI analytics, help in creating personalized conditioning programs that keep athletes in peak form while minimizing the risk of injury.

Injury prevention extends beyond physical conditioning. AI tools can now analyze players' movements to predict and prevent potential injuries. Amateur teams, often without access to full-time medical staff, benefit immensely from these predictive insights. By identifying early signs of stress or strain, AI systems alert coaches to modify training loads, or take preventative action, ensuring that players remain healthy throughout the season.

AI is not just a tool for players and coaches but also enhances the fan experience in amateur leagues. Interactive platforms and virtual assistants engage fans by delivering live game updates, player statistics, and tailored content. This heightened level of engagement transforms community supporters into informed fans who feel more connected to the game and the players. These digital tools can simulate professional-level commentary and analysis, making amateur matches more exciting to follow.

The integration of AI also facilitates efficient league administration. Scheduling matches, managing teams, tracking player registrations, and even handling league communications can be streamlined using AI-driven management systems. This automation frees up valuable time for league organizers and coaches, allowing them to focus more on game development and less on administrative tasks.

Community initiatives are also enriched by AI. Leagues can leverage technology to promote inclusivity and participation, ensuring that soccer becomes a unifying activity. AI can identify areas or demographics with lower participation rates, prompting targeted outreach programs. By understanding community dynamics through data, leagues can implement initiatives that increase accessibility and engagement.

The socio-economic impact of AI in amateur leagues is profound. By making advanced training and analytical tools accessible to everyone, AI bridges the gap between amateur and professional levels. Talented players from underprivileged backgrounds now have opportunities to develop their skills and showcase their talent. This democratization of technology ensures that soccer can be a viable career path for anyone, regardless of their socio-economic status.

Looking ahead, the future of amateur soccer looks brighter with AI. The technology continues to evolve, making tools more intuitive and affordable. Emerging technologies like augmented reality and virtual reality are likely to be integrated into training regimes, offering immersive learning experiences. With AI, the dream of every amateur player is within reach—not just playing for fun but realizing their potential to become professionals.

To sum up, AI is revolutionizing amateur leagues by providing tools that enhance performance, improve training, prevent injuries, and engage fans. It democratizes access to advanced technology, offering equal opportunities for talent discovery and development. As AI continues to refine its algorithms and applications, the boundary between amateur and professional soccer will become increasingly blurred, fostering a new era where talent truly knows no limits.

community initiatives

Grassroots soccer is the heartbeat of the sport, where young talents emerge and community spirit thrives. The use of artificial intelligence (AI) in grassroots initiatives isn't just a technological leap; it's a social revolution. AI bridges the gap between amateur enthusiasts and professional capabilities, creating a platform that nurtures talent, promotes inclusivity, and fosters a sense of belonging.

One of the most remarkable aspects of AI in grassroots soccer is its ability to democratize access to advanced training tools. Traditionally, sophisticated coaching techniques and performance analysis were exclusive to professional clubs. Now, AI-powered apps and platforms like Techne Futbol and DribbleUp are readily available to anyone with a smartphone. These technologies provide individualized training regimes, feedback, and progress tracking, helping young players improve their skills without the need for elite training facilities.

AI isn't just about enhancing individual performance; it's also playing a crucial role in uniting communities. Take, for instance, the creation of AI-driven platforms that organize local soccer leagues and tournaments. These platforms leverage AI to effectively manage team rosters, schedules, and even predict weather disruptions. Such efficient organisation not only boosts participation but also builds local bonds, making soccer more accessible and enjoyable for everyone.

Furthermore, AI facilitates talent discovery in ways that were previously unimaginable. AI-driven scouting tools can analyze countless hours of gameplay footage, identifying standout talent in various community leagues. These tools don't just spotlight star players; they also help coaches understand each player's strengths and weaknesses, leading to more tailored and effective coaching strategies. This approach ensures that no promising talent goes unnoticed, offering a genuine meritocratic pathway to higher-level opportunities.

AI in grassroots soccer isn't limited to on-the-field enhancements alone. Off the field, AI is being leveraged to promote diversity and inclusion. Programs focused on reducing biases in talent identification are becoming more prevalent, ensuring that players from diverse backgrounds receive equal opportunities. AI algorithms, trained to disregard factors such as race, economic status, and gender, can offer a more objective analysis of a player's potential, fostering a more inclusive environment.

In many communities, AI-driven initiatives have also revolutionized the way people engage with the sport. Local clubs are utilizing AI to maintain their fields more efficiently, using predictive analytics to manage field conditions and avoid cancellations due to poor weather. Additionally, AI-powered scheduling systems can optimize the use of shared fields, ensuring that multiple teams can practice and play with fewer conflicts. This maximization of resources encourages more frequent play and better overall community engagement.

Community health and well-being are also benefiting from AI initiatives in grassroots soccer. Wearables and other AI-driven health monitoring tools are increasingly being adopted to track players' fitness levels and prevent injuries. This proactive approach not only assures player well-being but also fosters a culture of care and support within the community. Parents and coaches can rest assured knowing that young athletes are being monitored with the highest technological standards.

The outreach potential of AI in grassroots soccer can't be overstated. AI's ability to bring world-class soccer knowledge into the palm of one's hand means that even remote and underserved communities can have access to top-tier coaching and development programs. Through partnerships with non-profits and community organizations, AI can help bridge the gap between privileged and less privileged areas, fostering equal opportunity for all aspiring players.

Education and training are other areas where AI is making a significant impact. Workshops and seminars on AI's role in soccer are increasingly being conducted in local communities. These initiatives not only educate young players about the modern tools at their disposal but also inspire a new generation of tech-savvy athletes. By exposing kids to AI early, we prepare them for a future where technology and sports are deeply intertwined.

Moreover, AI is protecting the heritage and inclusivity of grassroots soccer by recording and analyzing matches, making local legends out of community players. Platforms dedicated to archiving community matches provide a historical record, preserving the contributions and stories of local heroes. This documentation fosters a sense of pride and continuity in communities, connecting generations through the beautiful game.

An inspiring example comes from initiatives like the Common Goal Movement, where clubs and organizations leverage AI to ensure funds and resources are optimally distributed to reach as many underprivileged soccer communities as possible. AI's precise data analytics ensure that every dollar and resource is used effectively, maximizing the social impact of every initiative.

Community initiatives also extend to enhancing spectator experiences. AI-driven platforms can create customized content for local fans, tailored to their interests and preferences. Whether it's through augmented reality, immersive match highlights, or community engagement features on social media, AI is making local soccer more exciting and engaging for everyone involved.

The transformative power of AI extends beyond the technical sphere, embedding itself into the very fabric of grassroots soccer communities. From training enhancements to social inclusion, from preserving local legacies to engaging fans in unprecedented ways, AI proves to be a catalyst for positive change. By making advanced tools accessible and fostering a sense of community, AI doesn't just elevate the game at the grassroots level; it redefines what's possible.

As we move forward, the potential for AI-powered community initiatives in grassroots soccer is boundless. By continuing to innovate and prioritize inclusivity and equal access, we can ensure that the beautiful game remains a source of joy, development, and unity for generations to come. Whether it's through advanced analytics, personalized training, or community-focused platforms, AI is setting down roots that promise to bloom into a brighter, more inclusive future for soccer.

Chapter 21:
Future Trends in AI and Soccer

As we look ahead, the intersection of AI and soccer promises advancements that could further revolutionize the sport in ways currently unimaginable. Emerging technologies like machine learning and neural networks will likely drive more accurate predictive insights, offering data-driven forecasts on player performance and game outcomes with ever-increasing precision. Enhanced virtual reality (VR) and augmented reality (AR) systems could transform training sessions and fan experiences alike, offering immersive, interactive environments that were once the stuff of science fiction. These innovations will not only refine tactical planning and injury prevention but also bridge the gap between professional and grassroots levels, democratizing access to high-quality training resources. By integrating these cutting-edge tools, the beautiful game stands on the brink of an era where strategic and operational decisions are both more informed and impactful, propelling soccer into a future limited only by our imagination.

Emerging Technologies

As we cultivate our understanding of artificial intelligence's role in soccer, it becomes clear that emerging technologies are poised to revolutionize the sport in unprecedented ways. From machine learning algorithms that predict player performance to sophisticated wearable devices that monitor physiological data, the landscape of soccer is changing rapidly. These technologies offer insights and efficiencies that were previously unimaginable, opening new avenues for enhancing every aspect of the game, from strategy formulation to player health and even fan engagement.

One particularly promising area of development is the use of *machine learning* algorithms to analyze vast amounts of data. These algorithms can process player statistics, in-game actions, and even external factors like weather conditions to offer predictive insights. Teams that integrate these machine learning models into their decision-making processes can optimize their strategies in real-time, giving them a competitive edge. For instance, a team might adjust its formation or substitute players based on the real-time analysis provided by these algorithms, thereby reacting dynamically to the ebb and flow of the game.

In the realm of **player performance monitoring**, wearable technology stands out as an exciting frontier. Modern wearables not only track physical metrics like heart rate and speed but also capture intricate data such as muscle fatigue and hydration levels. When coupled with AI, this data can be analyzed to make personalized training recommendations and even predict potential injuries. By understanding the physical limits and conditions of players through these smart devices, coaches can make informed decisions to prevent injuries and extend careers, ultimately preserving their investments in talent.

Moreover, advancements in *virtual reality (VR)* and *augmented reality (AR)* are making significant inroads into training regimens. Virtual coaching assistants can simulate real-world game scenarios, offering players a chance to practice and refine their skills in a controlled, data-rich environment. These simulations can replicate the conditions of upcoming matches, preparing players psychologically and physically. By tackling these virtual challenges, players can enhance their situational awareness and decision-making skills, translating to improved performance on the actual field.

Another intriguing development is the integration of AI into **video analysis tools**. Advanced video analysis can break down game footage frame by frame, identifying patterns, weaknesses, and opportunities. These tools allow coaching staff to provide detailed, actionable feedback to players. For example, a forward might receive insights on their positioning and timing, while a defender may understand space management better. By dissecting both their own and their opponents' performances, teams can train more efficiently and develop counter-strategies that are informed by empirical data.

In terms of *fan engagement*, AI-driven chatbots and personalized content are rewriting the script. Imagine a virtual assistant that knows your favorite team, player, and even your preferred types of matches. This AI can offer personalized match updates, interpret statistical data in layman's terms, and even simulate post-match discussions tailored to your interests. Such technology can deepen fans' emotional connection to the sport, making them feel valued and understood. Furthermore, interactive platforms powered by AI enable fans to participate in predictive games and fantasy leagues, keeping them engaged throughout the season.

The broadcast experience is also evolving, thanks to intelligent commentary systems and real-time statistics. AI can provide instant, insightful commentary that adapts to the flow of the game. It can highlight key moments, predict possible outcomes, and even offer historical context during live broadcasts. Real-time statistics provided on-screen can give viewers a better understanding of the game, from player metrics to team dynamics. This fusion of live action and data analytics makes the viewing experience richer and more informative.

Emerging technologies also touch on ethical considerations, particularly in areas of **data privacy** and **AI fairness**. As wearables and other monitoring devices collect data, ensuring this information is used ethically and stored securely becomes paramount. Likewise, AI systems must be free of bias to ensure fairness across all levels of the game—from talent scouting to gameplay analysis. Addressing these concerns while maximizing the benefits of these technologies will be one of the essential balancing acts for the future of AI in soccer.

Looking further ahead, we anticipate that AI will become even more integrated into the tactical and operational aspects of soccer. Emerging technologies like *neural networks* are already being tested to predict game outcomes with remarkable accuracy. These complex algorithms mimic human brain processes, making them capable of learning and improving over time. As these technologies mature, teams will have even more powerful tools at their disposal, capable of offering deeper insights and more sophisticated predictions.

Startups and tech giants alike are also investing heavily in developing new AI applications tailored specifically for soccer. From AI-driven camera systems that automatically capture and tag game events to sophisticated software that can simulate millions of game scenarios, the breadth and depth of ongoing research are staggering. This investment demonstrates the massive potential and business opportunity seen in merging AI with soccer.

One fascinating speculative frontier is the use of **biometric data** and gene sequencing. Although still in its infancy, there is growing interest in how genetic information might influence talent identification and development. Imagine a future where AI analyzes genetic data to predict a player's potential, or even optimizes training programs based on individual genetic profiles. Such technologies, while controversial, could redefine our understanding of talent and performance in soccer.

As we continue to explore the possibilities, it's clear that the synergy between soccer and technology is just beginning. The advancements already visible are merely the tip of the iceberg, hinting at a future where AI and emerging technologies could reshape every aspect of this beloved sport. As these innovations evolve, so too will the beautiful game, maintaining its timeless appeal while embracing the cutting-edge tools of the 21st century.

Predictive Insights for the Next Decade

The next decade in soccer promises to be revolutionary, driven by the accelerating momentum of artificial intelligence. As AI continues to evolve, its role within the realm of soccer is becoming indispensable. Predictive analytics, a subset of AI, stands at the forefront of this evolution, offering profound insights that will shape the future of the game. We're not merely contemplating marginal gains but rather transformative changes with predictive insights delivering a deeper understanding of play patterns, injury risks, and even fan behavior.

Predictive analytics will be a game-changer in the tactical aspect of soccer. Coaches and analysts will harness AI-driven models to foresee the most effective strategies against specific opponents. These models will go beyond basic statistical data, incorporating player fatigue levels, weather conditions, and even psychological factors. Instead of relying on intuition alone, decision-makers can turn to nuanced predictions that account for both macro and micro factors affecting performance. This will lead to more informed and agile tactical adjustments, potentially altering the dynamics of matches in real-time.

Injury prevention will benefit immensely from predictive capabilities. As wearable technology becomes more advanced and widespread, the data gathered can offer unparalleled insights into players' physical conditions. Predictive models will identify signs of potential injuries before they occur, allowing for preemptive interventions. By analyzing historical data and patterns, these insights will spearhead personalized training and recovery programs tailored to individual athletes, thus extending careers and enhancing overall team performance.

Player development is another area poised for radical transformation. Predictive analytics can identify budding talent at an earlier stage, enabling clubs to invest in potential stars before their market value skyrockets. Youth academies will utilize AI-driven scouting systems that are far more accurate and less biased than traditional methods. These systems will analyze a plethora of data points from on-field performance to off-field habits, providing a comprehensive profile of young athletes, ensuring that no talent goes unnoticed.

Moreover, predictive insights will elevate the fan experience to new heights. AI will personalize content delivery based on individual preferences, creating a more immersive and engaging viewing experience. Fans will receive real-time predictions on match outcomes, player performance, and other game-related events, turning passive spectators into actively engaged participants. Personalized recommendations for merchandise, fantasy leagues, and interactive features will further deepen the connection between fans and the sport they love.

Broadcasting will also evolve with predictive insights playing a crucial role. Real-time AI-driven predictions during broadcasts will offer audiences a richer, more informative viewing experience. Commentators will have access to predictive models that provide insights into potential game-changing moments, enabling them to offer more nuanced and informed analysis. This will not only enhance viewer engagement but also attract new audiences who appreciate the enhanced depth and detail provided by predictive analytics.

The role of AI in contractual negotiations cannot be overstated. Predictive models can evaluate player performance over time, considering various factors such as age, injury history, and potential for future growth. This will enable clubs and agents to enter negotiations with a more data-driven approach, ensuring fair and accurate valuation of players. The financial aspects of soccer will become more transparent and equitable, driven by insights that predict players' future contributions and market value.

Furthermore, predictive insights will play a pivotal role in addressing ethical and social considerations within the sport. By identifying patterns of inequality or bias in scouting and recruitment, predictive models can aid in creating a more inclusive and diverse player base. This will ensure that opportunities are based on talent and potential rather than subjective judgments, leading to a fairer and more equitable landscape in soccer.

AI's predictive capabilities will also support mental health initiatives for players. By analyzing behavioral patterns and various psychological indicators, AI can offer early warnings of potential mental health issues. Personalized mental health support programs can be developed based on these insights, offering preventive care and timely interventions. This proactive approach will significantly improve players' well-being, both on and off the field.

In grassroots soccer, predictive analytics will democratize access to advanced insights typically reserved for elite levels. Local clubs and communities will be able to harness these technologies, leveling the playing field and fostering talent development from the ground up. AI will offer smaller clubs the ability to train and develop players with the same precision and effectiveness as top-tier teams, transforming the overall quality of the sport at all levels.

As AI technology evolves, its integration with big data will further amplify its predictive power. Teams that adeptly combine vast datasets with sophisticated AI models will gain a competitive edge. By filtering through enormous amounts of data to highlight actionable insights, AI will allow for more strategic decisions in both on-field performance and off-field management. Teams that embrace these innovations will not only lead in performance but also in the business aspects of soccer.

Over the next decade, the role of AI in soccer will merge even more seamlessly with human expertise, creating a new paradigm where technology complements intuition and experience. Predictive analytics will be at the heart of this synergy, offering insights that enhance every facet of the game. Coaches, players, analysts, and fans alike will benefit from these advancements, making soccer not only a beautiful game but a smarter one.

In conclusion, the predictive insights driven by AI will revolutionize soccer in ways we can only begin to imagine. The next decade will see the sport becoming more strategic, player-centric, and inclusive, powered by data and algorithms that anticipate the future. As we stand on the cusp of this transformation, it's clear that AI's potential to predict, enhance, and inspire is boundless, promising a thrilling new era for soccer.

Chapter 22:
Building an AI-Ready
Soccer Organization

To thrive in the modern world of soccer, clubs must evolve into AI-ready organizations. This involves comprehensive structural changes that integrate advanced technology seamlessly into everyday operations. From retraining staff to embracing new paradigms, the journey requires considerable effort and dedication. Coaching teams need to develop a deep understanding of AI tools, enabling them to personalize training programs and optimize player performance. Meanwhile, players must become adept at using wearable tech and engaging with AI-driven feedback for continuous improvement. By building an AI-ready organization, clubs not only enhance their competitive edge but also pave the way for a more data-driven, efficient, and adaptive approach to the beautiful game.

Structural Changes

Implementing artificial intelligence in soccer isn't just about introducing new technologies; it requires a fundamental overhaul of an organization's infrastructure, processes, and mindset. As we delve into the structural changes necessary for building an AI-ready soccer organization, it's crucial to understand that these modifications won't happen overnight. They're part of a long-term strategy that necessitates commitment from all levels of the organization, from top executives to ground-level staff.

First and foremost, leadership must be fully committed to the idea of evolving into an AI-ready entity. This involves not only financial investment but also a cultural shift. Executives and decision-makers need to champion the vision and communicate it persuasively to everyone involved. They should be prepared to confront and address resistance, which is inevitable when dealing with sweeping changes. The shift to an AI-centric organization is as much about altering mindsets as it is about upgrading technology.

Central to this transformation is the establishment of a dedicated AI team. This group should consist of a mix of data scientists, machine learning experts, and software engineers who specialize in AI applications for sports. Collaboration is key. This team must work closely with coaches, medical staff, and performance analysts to tailor AI solutions that address the unique needs of the soccer world. For instance, data scientists can develop algorithms to analyze player performance, but these tools will only be effective if they incorporate insights from coaches who understand the nuances of the game.

Creating a data-first environment is another critical step. Traditional soccer organizations often rely on intuition and experience to make decisions. While these elements will always have a place in the sport, incorporating data-driven insights can significantly elevate decision-making processes. To do this, clubs need to invest in data collection and analysis infrastructure. This includes not just the hardware, like sensors and cameras, but also the software and cloud services that enable the storage and analysis of vast amounts of data.

Integrating AI into the coaching and training staff is also essential. Coaches and trainers need to be educated about the benefits of AI and trained to use new technologies effectively. They should feel empowered rather than threatened by these advancements. This can be achieved through workshops, seminars, and ongoing training programs. Establishing a feedback loop where coaches can provide input on the AI tools will also ensure that these technologies are continually refined and adapted to better serve the team's needs.

Beyond the team-specific considerations, the organizational structure has to be adaptable to rapid technological advancements. The sports industry evolves quickly, and AI technologies can become obsolete almost as soon as they are implemented. Thus, the organizational framework should allow for flexibility and innovation. An iterative, agile approach can be beneficial here, where small-scale projects are tested, evaluated, and scaled based on their success.

Communication channels within the club must be optimized for the AI era. Data-driven insights should flow freely and reach the right stakeholders without delay. This might mean setting up new lines of communication between the AI team and the coaching staff, or it could involve adopting new software platforms that facilitate better information sharing. Clear, quick, and effective communication is a cornerstone of any AI-ready organization.

Moreover, partnerships and collaborations with technology providers, academic institutions, and other pioneering clubs can be highly beneficial. Such alliances can bring in external expertise, foster innovation, and share the financial burden of large-scale AI projects. For example, working with a university's computer science department could provide the club access to cutting-edge research and a pool of talented students eager to apply their knowledge to real-world problems.

On the player side, athletes must be brought into the fold. They should understand how AI will be used to enhance their performance, improve their training, and prevent injuries. Transparency is key here to ensure buy-in from players who might otherwise be skeptical of these new technologies. By involving them in the process and showing tangible benefits, clubs can alleviate fears and foster an atmosphere of cooperation and mutual benefit.

The role of data privacy and ethics cannot be ignored in this transformation. With vast amounts of personal data being collected from players, fans, and operations, organizations must establish robust data protection protocols. Ethical considerations, such as consent and data ownership, should be addressed comprehensively. These measures build trust and ensure that the organization remains compliant with regulations while maintaining a moral high ground.

Finally, it's essential to recognize that the path to becoming an AI-ready soccer organization is iterative and ongoing. Regular assessments and updates should be part of the organizational strategy. This ensures that the club remains at the cutting edge of technology and doesn't fall behind. Continuous learning and adaptation will keep the organization resilient in the face of constant technological advancements.

By embracing these structural changes, soccer organizations can position themselves not just to survive but to thrive in an AI-powered future. The benefits of such transformation are far-reaching, influencing everything from game strategy to player welfare and fan engagement. As the intersection of technology and soccer continues to evolve, organizations willing to make these significant shifts will lead the way in modern sports. This proactive approach ensures that the beautiful game retains its charm while embracing the innovations that can propel it to new heights.

Training Staff and Players

In the realm of building an AI-ready soccer organization, training staff and players is a cornerstone. Transitioning to an AI-driven ecosystem requires not only cutting-edge technology but also a well-prepared team that understands how to leverage these tools. Ensuring that every member, from the coaching staff to the players, is well-versed in utilizing AI can yield significant dividends on and off the field.

Training staff members in AI begins with a comprehensive overview of what artificial intelligence can achieve in the context of soccer. Coaches and analysts need to understand the fundamentals of AI, including machine learning, neural networks, and data analytics. This foundational knowledge empowers them to interpret AI-generated insights and integrate them into their coaching strategies. A robust training program typically includes workshops, seminars, and hands-on sessions led by AI experts and tech professionals who specialize in sports analytics.

Beyond the basics, staff need to delve into specific AI applications relevant to their roles. For instance, coaches might focus on using AI for tactical adjustments and real-time decision-making. Analysts, on the other hand, may dive deeper into data visualization tools and predictive models that assist in player development and performance monitoring. Equally essential is instilling an understanding of the limitations of AI, ensuring the staff can critically evaluate AI recommendations and cross-reference them with their own expertise.

Players, while not needing as detailed a grasp of AI as their coaches, must also be trained effectively. They should understand how wearable technologies track their movements and physiological metrics. This awareness helps players appreciate how data collection translates to personalized training regimens and injury prevention protocols. Workshops for players might focus on interpreting feedback from AI systems and incorporating suggestions to enhance their gameplay.

A collaborative culture where staff and players continuously communicate is vital. Connecting insights generated by AI to individual and team objectives fosters a unified approach to leveraging technology. Clubs that have successfully integrated AI often highlight the importance of this collaborative culture. Open lines of communication ensure that technological interventions are not seen as replacements for human intuition but as enhancements to it.

Another critical component is the continuous education needed to keep up with advancements in AI technology. As AI evolves, so too must the knowledge and skills of staff and players. Regularly updating training programs to reflect the latest in AI applications in sports keeps the organization at the cutting edge. Partnerships with academic institutions and tech companies can facilitate this ongoing education, bringing fresh perspectives and technological innovations to the team.

The integration of AI in training routines can be gradual but strategic. Starting small with pilot projects can help ease the transition. For instance, beginning with AI-driven video analysis tools for post-match reviews can showcase the immediate benefits of technology. As the staff becomes more comfortable and adept, more complex systems like real-time analytics during matches can be introduced.

It's also important to consider the psychological aspect of incorporating AI into everyday training. Both staff and players might initially feel a sense of apprehension about being evaluated by machines. Creating a supportive environment that emphasizes AI as a complement to human skills rather than a threat can mitigate these concerns. Highlighting success stories within and outside the organization where AI contributed to significant performance improvements can be particularly motivational.

Customization of AI training programs is another aspect that can't be overlooked. While the core principles of AI remain the same, different players and staff roles require tailored approaches. Personalized training sessions that cater to individual learning curves and areas of application ensure that everyone gets the most out of the technology. This customization acknowledges that not everyone will interact with AI in the same way, and educational efforts should reflect this diversity.

Finally, fostering an environment that encourages experimentation and feedback is crucial. AI itself is built on iterative processes; hence, the organizational culture should mirror this continuous improvement ethos. Staff and players should feel empowered to try new AI tools, provide feedback, and suggest modifications. This participatory approach not only improves the effectiveness of AI applications but also fosters a sense of ownership and accountability.

In conclusion, training staff and players is not just about teaching them to use AI tools but also about building a culture of collaboration, continuous learning, and adaptation. Successful integration of AI into a soccer organization requires a multi-faceted training approach, structured yet flexible, that acknowledges both the technological and human aspects of the game. It's about crafting a symbiotic relationship between AI and the people who use it, paving the way for a future where technology enhances every facet of soccer, from the training ground to the final whistle.

Chapter 23:
Risks and Limitations of AI in Soccer

As much as AI is transforming soccer, it's essential that we consider its risks and limitations. Over-reliance on technology can lead to a reduction in the intuitive and creative elements of the game that fans cherish. While AI tools can analyze vast amounts of data, they still struggle with the nuances and unpredictability that human judgment can provide. Additionally, AI systems often reflect the biases present in their training data, which can result in unfair decisions or overlooked talent, particularly concerning underrepresented groups. The implementation of AI requires a balanced approach, ensuring that the technology enhances rather than overshadows the human aspects of soccer.

Over-reliance on Technology

The appeal of artificial intelligence (AI) in soccer is undeniable. It promises to revolutionize the game, making it faster, smarter, and more efficient. However, the enchantment with technological prowess brings a risk that needs addressing: over-reliance on technology. While AI offers a treasure trove of data-driven insights, there's a distinct danger in abandoning the human elements that make soccer what it is.

Soccer, at its core, is a human endeavor. It's about the grit, emotion, and intuition that players and coaches bring to the field. Over-relying on AI could dilute these human aspects. If decisions are predominantly driven by algorithms, the unpredictable nature that fans love may start to vanish. A well-placed through-ball or that one-off incredible volley depends as much on a player's instinct and flair as on statistical odds.

There's also a cultural component to consider. Soccer traditions vary widely around the globe, and an over-dependency on technology might narrow these rich, diverse approaches into more homogenous strategies dictated by data. This risk threatens not just the global appeal of soccer but also the uniqueness of individual playing styles and tactical philosophies that have evolved over decades.

The psychological aspect is another matter. Players can become overly reliant on guidance from AI, undermining their ability to make quick, independent decisions on the pitch. This can stifle creativity and diminish the intuitive, split-second judgments that often make the difference between victory and defeat. Coaches, too, might find themselves trapped in a similar dependency loop, forsaking their innate game understanding for what the "numbers" suggest.

Moreover, there's a concern about the accessibility and democratization of AI technology. Top clubs with substantial financial resources are more likely to benefit from state-of-the-art AI systems, while smaller clubs could struggle to keep up. This creates a widening gap, reinforcing the divide between the elite and the underdogs—a scenario that goes against the very spirit of competition inherent in sports.

Let's also consider the technological limitations themselves. AI systems rely heavily on the quality and volume of data they are fed. Inaccurate or biased data can lead to flawed algorithms, which in turn can produce unreliable outcomes. Hence, an excessive dependence on tech might inadvertently prioritize flawed feedback mechanisms over human intuition and experience, leading to poor decision-making.

Critically, the excitement around AI might also outpace the development of sufficient regulatory frameworks to ensure its ethical use. Soccer, like any other sport, thrives on its integrity and fair play. Over-reliance on AI without proper oversight could open the door to manipulations that could tarnish the game's reputation and fairness.

One must ponder another sobering possibility: what happens when technology fails? Machines break, algorithms glitch, and software bugs happen. How prepared are teams to depend entirely on systems that are fallible? If a software bug occurs during a critical game analysis, the ensuing chaos could be detrimental to the team's strategy and morale.

As we delve into the realms of AI-driven analytics and insights, it's imperative to strike a balance. Technology should serve as an enhancement, not a crutch. Coaches and players should use AI tools as supplementary aids rather than replacements for traditional coaching and experience-based tactics. This hybrid approach ensures that human ingenuity isn't overshadowed by technological determinism.

Furthermore, fostering an environment that values continuous learning and adaptation will help mitigate risks. Coaches and players must stay educated about both the potentials and pitfalls of AI. By understanding the limitations and ethical aspects of technology, teams can make informed decisions, using AI to bolster their strategy without becoming enslaved to it.

Equally essential is fostering a collaborative approach where AI is one of many tools in a diverse arsenal. Combining AI insights with traditional methods can offer a multi-faceted perspective that leverages both technology and human judgment. This integrated approach will enable teams to adapt to various on-field situations effectively.

In conclusion, while AI holds considerable promise in transforming soccer, a balanced approach is crucial to prevent over-reliance. By blending the best of technology with human intuition and creativity, soccer can retain its soul while embracing innovation. This equilibrium will ensure that AI serves as a powerful ally rather than an overruling master, preserving the essence and unpredictability that make soccer the beautiful game it is.

Addressing AI Bias

Artificial Intelligence (AI) has been integrated into soccer with great enthusiasm. While it's a powerful tool for enhancing performance, scouting talent, and even fan engagement, it's crucial to acknowledge that AI comes with its own set of biases. Bias in AI isn't just a technical glitch—it's a serious issue that can skew results, perpetuate existing inequalities, and even create new ones.

Let's begin by understanding what we mean by AI bias. Essentially, bias in AI systems stems from the data they are trained on. If the training data is imbalanced or contains historical biases, the AI system will inevitably reflect those biases in its predictions and decisions. For instance, if an AI model is trained predominantly on player data from European leagues, it might undervalue talent from less represented regions. This asymmetry can lead to unequal opportunities for players and potentially distort the global talent pool.

One area where AI bias can manifest is in player scouting and recruitment. Algorithms designed to identify promising talent may inadvertently favor players who fit traditional molds, thus overlooking outliers who could revolutionize the game. For example, Lionel Messi was considered too small and frail during his early years. A biased AI system might have overlooked him, depriving soccer of one of its greatest talents. It's essential for AI models to be diverse, not just in the data they consider but also in the parameters they evaluate.

Moreover, AI bias can have significant implications for game strategy and in-game decisions. Coaches relying heavily on AI for tactical adjustments might find their options constrained by the AI's inherent biases. If the system has a bias towards certain styles of play, it could lead teams to adopt less effective strategies against certain opponents. This isn't just a theoretical concern; it's a practical issue that can affect the outcomes of matches and the careers of coaches and players alike.

Real-time data analytics, powered by AI, can also fall prey to biases. Systems analyzing player movements and game stats might prioritize metrics that align with traditional understandings of performance. These biases can influence decisions made on the field, from substitutions to game tactics. If not properly addressed, AI could reinforce a narrow view of what makes a player or team effective, thereby stifacing innovation and creativity in the sport.

Bias in AI isn't limited to the pitch. It extends to training and player development, areas where personalized programs promise to revolutionize soccer. Personalized training plans, guided by AI, can help players optimize their performance. However, if the AI's recommendations are based on biased data, they might favor certain types of players over others. For example, players from underrepresented backgrounds may not receive training programs optimized for their unique needs and strengths, thus perpetuating existing disparities.

The impact of AI bias on fan engagement and experience is another area worth exploring. AI-driven platforms that offer personalized content and interactive experiences need to be designed ethically. If not, they might unintentionally marginalize certain fan demographics. For instance, AI models might prioritize content for fans based on geographic or socio-economic factors, thus excluding underrepresented groups. This could widen the gap between different fan communities, undermining one of soccer's greatest strengths: its universality.

So, how do we tackle AI bias in soccer? It starts with recognizing the problem and undertaking deliberate measures to mitigate it. First, it's critical to ensure diversity in training data. Data should encompass a wide range of leagues, player demographics, and playing styles. By diversifying the input data, we can help ensure that AI models are less likely to reproduce existing biases.

Next, transparency in AI algorithms is essential. While AI systems are often viewed as "black boxes," revealing the criteria and factors they consider can help identify and correct biases. This transparency extends to the entire lifecycle of AI development, from data collection to model training to deployment. Ethical guidelines and standards can play an essential role here, ensuring that AI systems align with broader social values and goals.

Another approach is involving human oversight in AI systems. While AI can process vast amounts of data quickly and efficiently, human judgment is necessary to interpret its outputs and override them when needed. Coaches, analysts, and other stakeholders should have the ability to scrutinize AI-driven insights and make adjustments based on their expertise and intuition. This collaborative approach can help ensure that AI augments human decision-making rather than dictating it.

It's also worth investing in continuous evaluation and feedback mechanisms. AI models should be regularly assessed for biases and their impact on decision-making. Feedback loops can help identify areas where AI systems are falling short, enabling ongoing improvements. This iterative process is key to ensuring that AI remains a tool for progress rather than a source of perpetuated inequities.

Education plays a pivotal role in addressing AI bias. Stakeholders across the soccer ecosystem, from coaches to administrators to fans, need to understand how AI works and the potential biases it can harbor. Training programs and workshops can equip individuals with the knowledge to identify, question, and address AI biases. By fostering a culture of awareness and critical thinking, we can collectively work towards more equitable AI applications.

Bias in AI also has a legal and ethical dimension. Regulatory frameworks and standards need to be established to govern the use of AI in soccer. These frameworks should mandate fairness, transparency, and accountability in AI systems. Additionally, ethical considerations should be embedded into the AI development process, ensuring that AI applications align with broader social and moral values.

Furthermore, addressing AI bias requires a multi-stakeholder approach. Collaboration between tech developers, soccer governing bodies, clubs, players, and fans is vital. Each stakeholder brings unique perspectives and expertise, contributing to a more holistic understanding of biases and how to mitigate them. This collaborative approach also enhances the legitimacy and acceptance of AI applications in soccer.

In conclusion, while AI holds immense potential for transforming soccer, it's imperative to address the biases it can introduce. By recognizing and proactively tackling these biases, we can ensure that AI serves as a force for good in the sport. Through diverse data, transparent algorithms, human oversight, continuous evaluation, education, ethical frameworks, and multi-stakeholder collaboration, we can harness the power of AI to create a more inclusive, equitable, and innovative soccer ecosystem.

Chapter 24:
Global Perspectives on AI in Soccer

A cross the globe, the adoption of artificial intelligence in soccer varies significantly, reflecting the diverse cultural and strategic priorities of different regions. In Europe, elite clubs leverage AI to gain competitive edges, using it for everything from advanced analytics to player health monitoring. Meanwhile, in South America, AI is increasingly being used to scout emerging talent, ensuring that the rich vein of footballing skill is continually tapped. In Asia, a tech-savvy fanbase drives the demand for AI-enhanced spectator experiences and interactive platforms, blending traditional fervor with modern innovation. Africa sees AI as a tool for leveling the playing field, offering grassroots initiatives the means to compete on a larger stage. As each region adopts AI in unique ways, it's clear that while the technology is universal, its application must respect and respond to local nuances and needs, driving the beautiful game forward in a way that feels authentic and inclusive.

Adoption in Different Regions

Across the globe, soccer's adoption of artificial intelligence varies greatly, shaped by unique regional characteristics, economic conditions, and cultural affinities. Europe's soccer powerhouses have been the forerunners in this technological arms race. Clubs like Manchester City, Barcelona, and Bayern Munich leverage cutting-edge AI to stay ahead of adversaries. Data analytics and machine learning algorithms inform everything from player recruitment to in-game tactics.

In England, the Premier League has embraced AI-driven solutions with an almost evangelical zeal. The league's financial muscle allows teams to invest in sophisticated data-tracking systems like those provided by STATSports and Catapult. These systems offer granular insights into player movements, fatigue levels, and even emotional states during games. The focus isn't just on gathering data but on making it actionable. Machine learning algorithms evaluate various scenarios, providing coaches with real-time recommendations for substitutions or tactical changes that could turn the tide of a match.

Germany's Bundesliga, known for its engineering precision, has married this trait with AI technologies. German clubs often collaborate with local universities and tech firms to develop bespoke tools that address specific tactical needs. Take the case of TSG Hoffenheim, a club that has utilized AI-driven video analysis tools to dissect opponents' strategies down to individual player tendencies. Such analytics have become a cornerstone for managing not only in-game decisions but also training routines tailored to exploit the weaknesses of forthcoming opponents.

By contrast, in Southern Europe, AI adoption has been somewhat tempered by cultural factors and economic constraints. Italian clubs, particularly in Serie A, have shown a more conservative approach. The focus remains heavily on traditional scout-based recruitment and game management methods. Nevertheless, exceptions exist. Juventus, for example, has begun to flirt with AI for tactical planning and player development. The club's collaboration with IBM to use the Watson supercomputer for analyzing vast amounts of game data is a significant step toward integrating AI into the Italian soccer ethos.

The picturesque but fiercely competitive leagues of Spain have seen varied levels of AI adoption. Barcelona, always an innovator in playing style, has expanded its horizon to AI-related analytics. Their state-of-the-art facilities include the Barça Innovation Hub, which functions as a living lab for testing AI and other emerging technologies. Real Madrid, another giant in the region, uses AI-driven insights for both player performance and fan engagement, creating a more interactive and immersive experience for their global fanbase.

Meanwhile, across the Atlantic, Major League Soccer (MLS) in the United States has taken a strategic approach toward AI, balancing between enhancing game performance and elevating fan engagement. The MLS has a younger fan demographic that is generally more tech-savvy, making AI solutions like virtual reality experiences and chatbot-driven fan interactions particularly appealing.

The league's centralized structure aids in the uniform adoption of AI-backed technologies. Players across different franchises are assessed using standard metrics, enabling a more equitable and comprehensive development system. Clubs like LA Galaxy and New York City FC are early adopters, leveraging AI for scouting talent both domestically and internationally. The technology is increasingly important as MLS looks to shed its "retirement league" image and become a breeding ground for global soccer talent.

In South America, soccer is virtually a religion, with passion often outweighing financial muscle. However, economic disparities have meant slower adoption of AI technologies compared to Europe and North America. Brazil and Argentina, blessed with rich soccer traditions, are starting to embrace AI, albeit at a slower pace. Clubs like Flamengo and Boca Juniors have begun dabbling with AI-powered performance analytics, but widespread adoption remains stymied by financial limitations.

Africa, often referred to as the "Sleeping Giant" of global soccer, presents a different story. Countries like Nigeria, Ghana, and South Africa show immense potential but struggle with resource constraints.

Here, AI adoption is driven more by national federations than by individual clubs. Federations are partnering with international organizations to bring AI-driven talent identification programs to youth academies. While the journey is long, there is a concerted effort to lay a technological foundation for the future.

In Asia, the story is one of contrasts. Japan and South Korea lead the region in integrating advanced technologies into their soccer infrastructure. Japanese clubs such as Vissel Kobe have enlisted the help of AI to understand player biomechanics, optimizing training regimens to reduce injury risks and enhance performance. South Korea's league is not far behind, embracing video analysis tools and predictive analytics to sharpen tactical awareness.

China, with its overarching goal of becoming a soccer powerhouse by 2050, has shown significant interest in AI. Supported by state-run initiatives and substantial financial backing, Chinese Super League clubs are investing in AI technologies at an unprecedented scale. The goal is not just to elevate domestic competition but to develop a pipeline of talent capable of competing on the world stage.

Both the Middle East and Australia represent rapidly growing markets in soccer AI adoption. In the Middle East, nations like Qatar and the UAE are leveraging their considerable financial resources to incorporate state-of-the-art AI technologies, ranging from game performance analytics to AI-driven fan engagement platforms. These investments are part of broader strategies to establish themselves as global hubs for sports excellence.

Australia's A-League, meanwhile, takes a holistic approach toward AI integration. Here, technology serves not only to enhance performance but also to cultivate a soccer culture that's intrinsically linked to the community. AI tools are used extensively in youth development programs and community initiatives, emphasizing long-term developmental gains over immediate competitive successes.

Despite these geographic disparities, one thing is clear: AI in soccer isn't a fleeting trend but a transformative force reshaping the landscape of the sport. From the glittering stadiums of Europe to the burgeoning leagues of Africa and Asia, the embrace of AI varies but moves inexorably forward. The future will see even more regions jumping on the bandwagon, driven by success stories and the undeniable advantages that artificial intelligence offers to the beautiful game.

This global mosaic of AI adoption becomes a testament to the sport's universal appeal and its ability to adapt and thrive in an ever-changing technological landscape. As regions learn from each other and share breakthroughs, soccer's AI-driven evolution is set to accelerate, bringing new dimensions to a sport loved by billions around the world.

Cultural Impacts

The integration of artificial intelligence (AI) in soccer is more than just a technological upgrade; it's a cultural revolution reshaping how the sport is perceived, played, and celebrated globally. At the heart of this transformation lies a confluence of traditions and innovations, where cultures with deep-rooted soccer histories meet cutting-edge advancements.

Take, for instance, Europe, often considered the epicenter of soccer's rich legacy. Here, AI has been embraced with a nuanced approach, melding age-old fan rituals with new-age technology. In countries like England, Italy, and Spain, AI tools are widely used to analyze player performance and strategize game plans. Yet, these nations face the challenge of ensuring that the essence of the sport remains unchanged amidst burgeoning technological interventions. The art of scouting, for example, has evolved; the legendary "eye test" by seasoned scouts now stands alongside automated talent identification systems, adding a layer of precision to the age-old practice.

Across the Atlantic, AI's role in soccer expresses itself differently. In the United States, where Major League Soccer (MLS) has gained exponential popularity, AI has become a core component of the sport's rapid growth. Here, the emphasis is on enhancing the fan experience through personalized content and interactive platforms, a nod to America's affinity for technological integration in sports. AI-driven data analytics also play a significant role in attracting and training young talent, which aligns well with the American spirit of innovation and progress.

In South America, soccer is more than a sport—it's a way of life infused with passion and pride. The region's nations, like Brazil and Argentina, have historically produced some of the world's most phenomenal talent. The introduction of AI in these soccer-crazy cultures has been viewed with a mix of skepticism and excitement. Here, the challenge is balancing the fervor and creativity inherent in their style of play with data-driven approaches that might otherwise seem restrictive. Nevertheless, AI's potential in injury prevention and performance monitoring is quickly gaining acceptance, promising to protect and extend the careers of beloved stars.

Africa, often considered the cradle of raw soccer talent, presents another unique cultural landscape. Many countries on the continent have traditionally faced barriers like inadequate infrastructure and funding. However, AI offers a pathway to overcome these obstacles through cost-effective, scalable solutions. AI-driven scouting tools, for example, can identify talent in remote areas that might otherwise go unnoticed. As AI becomes more accessible, enthusiastic young African players are increasingly using technology to refine their skills, offering a modern complement to their natural abilities.

Moving to Asia, we see another diverse and evolving soccer culture. Nations like Japan and South Korea have been quick to adopt AI, aligning well with their broader technological advancements. In these countries, AI has been seamlessly integrated into youth training programs and professional leagues alike. Particularly interesting is China's aggressive approach to becoming a global soccer powerhouse, heavily investing in AI to train young athletes and analyze competitive strategies. This drive is not only about excelling in the sport but also about national pride and global reputation.

However, the cultural impact of AI in soccer is not solely confined to how the game is played or managed. It also touches deeply on the fan experience, altering the way spectators interact with the sport. In regions where soccer is almost a religion, AI-powered platforms have opened new avenues for fan engagement. Whether it's interactive apps that offer real-time statistics, chatbots providing personalized fan experiences, or virtual assistants that simulate historical matches, AI has brought fans closer to the game than ever before. This engagement fosters a sense of community that transcends physical boundaries, making the global soccer fanbase even more interconnected.

Language and regional colloquialisms also shape how AI tools are developed and adopted. For example, voice-activated AI systems in countries like Brazil need to understand Portuguese nuances and slang to appear user-friendly and authentic. Developers increasingly work with local experts to integrate these linguistic subtleties, ensuring that AI tools resonate culturally with their users. This localization extends to marketing strategies as well, where campaigns often fuse traditional soccer imagery with futuristic AI concepts to appeal to various cultural contexts.

Nevertheless, the cultural shift driven by AI is not without its controversies. Traditionalists in many regions worry that the increasing quantification of soccer might strip the game of its spontaneity and emotional depth. There's also the question of data privacy and ethics, particularly in societies with differing views on technology and surveillance. Fans and players alike express concerns over who owns the data and how it's being used. These issues necessitate ongoing dialogues and policies that respect both technological advancement and cultural sensitivities.

Moreover, it's essential to consider the socioeconomic impact of AI in soccer. In wealthier countries, clubs can afford cutting-edge AI tools that significantly enhance their performance and fan engagement capabilities. Conversely, in less affluent regions, there's a risk of widening the gap between elite and grassroots levels. Yet, AI also holds the promise of democratizing access to high-level coaching and performance analysis, provided there is a conscious effort to make these technologies accessible to all.

In conclusion, the cultural impacts of AI in soccer are multifaceted and deeply nuanced. The marriage of tradition and technology offers opportunities and challenges alike, reshaping how the sport is perceived, played, and enjoyed around the globe. As AI continues to evolve, it's crucial to strike a balance, ensuring that while we embrace innovation, we never lose sight of the cultural richness that makes soccer the beautiful game. The next chapters will delve deeper into specific instances of AI impact, illuminating further aspects of this fascinating transformation.

Chapter 25:
AI and the Beautiful Game:
A Balanced View

In the ever-evolving landscape of soccer, the integration of artificial intelligence offers a compelling glimpse into the future of the beautiful game. It's a dance between tradition and innovation, where cutting-edge technologies promise to elevate player performance, tactical strategies, and fan experiences to unprecedented heights. Yet, amid these advancements, it's crucial to reflect on what makes soccer so captivating: its unpredictability, its emotion, and its humanity. AI, with all its capabilities, should serve as a complement rather than a replacement. Ensuring this balance requires a mindful approach, where technology enhances rather than overshadows the human spirit that has always driven the sport. As we forge ahead, maintaining this equilibrium will be key to embracing the future without losing the soul of soccer.

Balancing Tradition and Innovation

Soccer holds a special place in the hearts of millions around the globe. Steeped in tradition, the sport has a cultural significance that's hard to quantify. Fans cherish the legends, the historic matches, and the timeless techniques that have been passed down through generations. At the same time, artificial intelligence (AI) has carved out a role that's too impactful to ignore, creating a dichotomy that the sport must navigate carefully.

Embracing AI in soccer doesn't mean abandoning centuries of tradition; it means finding a balance that reinforces the core values of the game while integrating modern advancements. This duality requires careful consideration from all stakeholders: players, coaches, club officials, and fans alike. Both realms—tradition and innovation—need not be mutually exclusive but can coexist to elevate the game to new heights.

The roots of soccer are intertwined with stories of grit, raw talent, and instinctual brilliance. Iconic matches and legendary players like Pelé, Maradona, and Zidane are celebrated not just for their statistics, but for the magic they brought to the field. Their unpredictable, human element is something AI can complement but never replicate. In the same breath, AI can amplify this magic by providing data-driven insights, identifying subtle patterns that even the keenest human eye might miss.

For instance, the intricate dribbling patterns of players or the stealth movements in a defensive line can be analyzed to an astounding degree of accuracy. This detailed analysis allows coaches to optimize strategies while maintaining the spirit of human creativity and flair that makes soccer the beautiful game.

From a coaching perspective, the traditional methodologies center around years of personal experience, intuition, and situational awareness. AI brings a systematic approach by offering real-time analytics and predictive modeling. Veteran coaches are now blending these AI-generated insights with their rich tapestry of experience, creating a more holistic coaching strategy. Think of AI not as a replacement but as an extension of the coach's expertise.

Cultural shifts within clubs also reflect this balance. Older club officials, who have lived through decades of changes, are increasingly open to AI but remain grounded in tradition. Newer generations embrace technology more readily, creating a balanced internal culture that respects the past while being excited about the future. This interplay ensures the club's ethos remains intact even as it evolves.

Fan experience is another domain where tradition and innovation blend seamlessly. Legacy fans may prefer the simplicity and nostalgia of a game unburdened by statistics, while younger audiences crave deeper engagement through immersive technologies. AI-driven platforms can deliver personalized content, from player stats to tailored highlights, addressing the varying preferences of a diverse fan base. This tailored approach respects the past by not forcing new technologies on those who might be reluctant while still offering cutting-edge engagement for tech-savvy fans.

The challenge lies in honing AI applications to enhance rather than overshadow the game's human elements. Innovations like real-time data during broadcasts and AI-generated tactical analysis during games should serve as tools that add depth to traditional narratives. When Cristiano Ronaldo scores a last-minute goal, it's not just about the data points but the emotion, the pressure, and the sheer artistry of the moment. AI can provide context without stripping away the sentiment.

From a player's perspective, traditional training drills focus on physical endurance, skill refinement, and teamwork. AI can enrich these training regimes by identifying areas for improvement down to the most minute detail. A striker might refine their positioning based on AI analysis of successful goal patterns, but it's their intuition, honed over years of traditional practice, that will ultimately guide their movements on the pitch.

The scouting and recruitment process also benefits from this balance. While AI systems can predict potential based on extensive data, the true essence of a player—heart, passion, and resilience—often reveals itself in moments that can't be quantified. Scouts who merge their instincts and expert judgment with AI insights create a more robust recruitment process. It's this dual approach that preserves the sanctity of player evaluations while harnessing the power of technology.

As soccer continues to evolve, striking the right balance between tradition and innovation will be crucial. The sport's rich history shouldn't be seen as a barrier to progress but as a foundation on which modern advancements build. Clubs worldwide are increasingly recognizing this, designing strategies that incorporate AI while staying true to soccer's core values.

Even as AI permeates nearly every facet of the sport, the essence of soccer remains unchanged. The roar of the crowd, the thrill of an unexpected goal, and the spirit of competition are timeless elements that AI will never replace. Instead, these can be enhanced through thoughtful integration, leading to a richer, more engaging experience for everyone involved.

Balancing tradition and innovation in soccer is not a zero-sum game. It's about leveraging AI to support what makes the sport universally beloved while ensuring that the human touch—the unpredictability, the emotion, the artistry—continues to shine. As we look to the future, the goal should be a harmonious blend that honors the past while embracing the transformative potential of AI. This balanced approach promises to keep the beautiful game both timeless and forward-looking.

Maintaining the Human Touch in Soccer

The rise of artificial intelligence in soccer brings transformative changes, impacting every aspect of the game from strategy to fan engagement. Yet, with this influx of technology, there's a crucial element that must not be overlooked: maintaining the human touch. Soccer isn't just a collection of data points and algorithms; it's a sport steeped in tradition, emotion, and human spirit. How do we balance these innovations with the essence of what makes soccer cherished by millions around the globe?

One of the most significant aspects of keeping the human touch in soccer lies in decision-making. AI can provide an array of data-driven insights to inform coaches and managers, but the final decisions on tactics and team selection must still come from humans. The intuition and experience that seasoned players and coaches bring to the table are invaluable. While algorithms can predict trends and outcomes, they cannot capture the nuances of a team's morale or a player's emotional state. These subtleties often influence the flow and outcome of a match.

Consider the role of a captain on the field. AI can track a player's performance metrics and stamina levels, but it cannot measure leadership qualities or the ability to inspire teammates during critical moments. These human qualities play a vital role in turning the tide during challenging matches. The decisions made by captains and leaders on the field aren't just based on statistics but rooted in years of experience, gut feelings, and split-second judgment calls that no machine can replicate.

Fan engagement offers another arena where the human touch remains indispensable. AI technologies can enhance the spectator experience through personalized content and interactive platforms. However, the passion and emotional connection that fans have with their favorite teams and players stem from human stories and relatable experiences. Social media, live interactions, and meeting players in person fuel this connection that data can only quantify but never truly capture.

While wearable technologies and AI-driven performance monitoring tools are remarkable, they should complement rather than replace the natural instincts of players. From amateur youth leagues to professional leagues, the joy derived from playing soccer isn't solely in the numbers. It's about the exhilarating experience of scoring a goal, the heartbreak of a missed penalty, and the bond formed between teammates. These emotional highs and lows create memorable moments that endear the game to fans worldwide.

Furthermore, training and player development must preserve a balance between technological advancements and traditional mentorship. Virtual coaching assistants and personalized training programs powered by AI can provide customized feedback and track improvements efficiently. Nevertheless, the guidance, encouragement, and sometimes tough love from a human coach are irreplaceable. A coach's ability to connect with players on a psychological level and to understand their unique motivations and fears fosters growth that AI cannot achieve alone.

It's essential to prioritize the narrative when it comes to storytelling within soccer. While AI can produce instant game reports and detailed analysis, it lacks the ability to embed a sense of emotion and context into those stories. Journalists and commentators play a pivotal role in creating compelling stories that resonate with fans, drawing from their understanding of the sport's history, culture, and emotional trends. They craft stories that build anticipation, celebrate victories, and mourn losses in ways that connect deeply with audiences.

Yet, there's an undeniable benefit of blending AI with human-driven insights. For example, analysts can use AI to pinpoint specific weakness in an opponent's strategy, but it takes human creativity and intuition to exploit these weaknesses effectively. A perfect blend of data analytics and human ingenuity enables teams to perform at their highest levels while retaining the spirit of the game.

Maintaining the human touch also entails ethical considerations. With AI making inroads into various facets of soccer, ensuring transparency, fairness, and respect for privacy is paramount. Decision-makers must involve various stakeholders, including players, coaches, fans, and ethics experts in discussions on how AI is employed. This collaborative approach helps to safeguard the human element and uphold the integrity of the sport.

Moreover, cultural and regional differences in how soccer is played and enjoyed globally must be respected. AI solutions should be adaptable to the diverse soccer cultures that exist. The essence of Brazilian flair, German precision, or the underdog spirit seen in smaller footballing nations, can't be distilled into algorithms. These cultural nuances bring richness to the game, making it universally beloved.

A well-rounded approach to adopting AI should emphasize continuous training and education for staff and players, ensuring they understand the technology but also know its limits. Encouraging mindful usage where tech aids but doesn't overshadow the human contributions can cultivate a balanced environment. Workshops, seminars, and training modules can highlight both the benefits and the boundaries of AI applications.

Human-error is an inseparable part of soccer that adds to its drama and unpredictability. While technologies like VAR (Video Assistant Referee) aim to reduce these errors, the spontaneous joy or outrage of a wrong call is a part of the game's emotional landscape. A certain level of imperfection is ingrained in the sport's charm, and completely eradicating it through technology could result in a less emotionally engaging experience.

Finally, the grassroots level of soccer plays a foundational role in maintaining the human touch. Community-driven initiatives, youth camps, and amateur leagues where passion overrides precision form the bedrock of future talent. AI can assist in these arenas, providing accessibility and resources, but the heart of these programs lies in community mentorship, involvement, and human connections.

The challenge is not just technological but philosophical: How can we adopt AI in soccer without losing its soul? The answer doesn't lie in opposing AI but in harmonizing it with the human elements that make soccer "the beautiful game." As we look ahead, the key is to see technology as an enabler, not a replacer. It should augment human effort, optimize potential, and highlight the narratives that make the sport extraordinary.

The true success of AI in soccer will be measured not just by the efficiency it brings but by how well it enhances these human experiences and preserves the traditions that have made soccer timeless. Finding this balance is our collective responsibility, ensuring that as we march towards an AI-enhanced future, the heart of soccer keeps beating with human passion.

Conclusion

The journey through the intricate interplay of artificial intelligence and soccer has been nothing short of enlightening for both enthusiasts and experts alike. By now, you must have a deep appreciation for how AI is shaping various facets of the beautiful game. From player development and team strategy to fan engagement and broadcasting, AI is not just an addition but a transformative force.

As the landscape of soccer continually evolves, AI emerges as a double-edged sword. On one side, it provides unparalleled opportunities to enhance performance, reduce injuries, and create more engaging experiences for fans. On the other, it brings forth ethical concerns, data privacy issues, and the potential for over-reliance. Yet, the balance of these forces is what will define the future of soccer.

One of the most impactful areas where AI has made strides is in game performance enhancement. Real-time data analysis and tactical adjustments are no longer the stuff of science fiction but are very much a reality. Coaches can make split-second decisions based on real-time insights, creating more dynamic and competitive matches.

Moreover, player performance monitoring has reached new heights. Wearable technology and advanced metrics allow for a comprehensive understanding of a player's physical and mental state. Predicting and preventing injuries through AI not only prolongs careers but also ensures player safety, making the game more humane.

The scouting and recruitment process has been revolutionized by AI too. Automated talent identification and predictive performance metrics mean that clubs can find the next star player more efficiently. It's no longer about who has the best eye for talent but who has the best algorithms and data.

In training and development, virtual coaching assistants and personalized training programs are preparing the next generation of soccer players. These technologies adapt and evolve based on individual needs, carving out tailored pathways for athlete improvement. It's a level of personalization that was unimaginable a decade ago.

Fan engagement has also been redefined in this digital age. Interactive platforms and personalized content draw fans closer to the action, regardless of their location. AI-driven chatbots and virtual fans create a new dimension of interaction, making the soccer experience more immersive than ever.

Broadcasting has not been left behind in this revolution. AI-powered intelligent commentary systems and real-time statistics offer viewers insights that enhance their understanding and enjoyment of the game. This move toward data-centric broadcasts caters to a generation that craves information at their fingertips.

Through our numerous case studies, we've seen success stories from top clubs that have leveraged AI to their advantage. These examples serve as blueprints for other clubs looking to adopt similar technologies. Despite the challenges, the rewards have proven to be monumental.

Ethics in AI remains a pivotal discussion. With data privacy concerns and debates over AI's role in sports, it's clear that governing bodies need to establish strong frameworks. Ensuring ethical practices will be crucial in gaining acceptance and trust from all stakeholders involved.

In the realm of women's soccer, AI has broken barriers and introduced technologies that are leveling the playing field. The case studies we explored demonstrate that AI is not gender-biased; it can uplift any league that chooses to embrace it, facilitating growth and development.

Youth development in soccer has also been supercharged by AI. Identifying talent early and nurturing them in AI-enhanced youth academies ensures a continuous pipeline of skilled players. This technology is vital in maintaining the sport's competitive edge globally.

The symbiosis between AI and big data has opened new avenues for team management. Integrating these technologies offers coaches and managers an all-encompassing view, enabling well-informed decisions that drive success on and off the pitch.

Financially, AI offers significant returns. From revenue generation to cost-benefit analyses, clubs can maximize their resources more effectively. This financial prudence is essential in an era where the economic landscape of sports is under constant scrutiny.

Even in player contract negotiations, data-driven valuation is changing the game. With AI, clubs and agents have a clearer picture of a player's worth, facilitating fair and objective negotiations. It's a move towards transparency and fairness in an otherwise opaque process.

The refereeing aspect of soccer has also seen a marked improvement. Technologies like VAR have minimized human error, and further advancements aim to make the beautiful game more fair and just. AI's role here cannot be understated as it ensures the integrity of the sport.

Mental health support for players is a burgeoning field where AI is making significant contributions. Monitoring mental well-being and offering AI-driven psychological support ensures that players can maintain their mental health alongside their physical fitness, creating a more holistic approach to athlete care.

AI's influence isn't limited to professional soccer alone. At the grassroots level, AI is transforming amateur leagues and community initiatives, democratizing access to advanced training methodologies. This grassroots revolution ensures that soccer's benefits reach every corner of the globe.

Looking ahead, the future trends in AI signify emerging technologies that promise to take the game to new heights. Predictive insights will offer foresight into the next decade, guiding clubs and players towards sustained success.

Building an AI-ready soccer organization involves structural changes and continuous training for staff and players. Embracing this digital shift is essential for staying competitive in an ever-evolving industry. It requires a collective effort from all parts of the organization.

Yet, it's essential to acknowledge the risks and limitations of AI in soccer. Over-reliance on technology can overshadow human ingenuity, and addressing AI bias remains a critical task. These challenges need careful management to harness AI's benefits fully.

Different regions adopt AI in soccer at varying paces, and the cultural impacts are significant. Global perspectives shed light on how technology is tailored to fit diverse cultural contexts, enriching the soccer experience worldwide.

Balancing tradition and innovation is the core mantra. Maintaining the human touch in soccer ensures that the essence of the game remains intact. AI should enhance, not overshadow, the human elements that make soccer a universal passion.

In essence, AI's role in soccer is a complex mosaic of opportunities and challenges. It has the potential to elevate every aspect of the sport, from grassroots development to elite performance. However, its integration must be handled wisely, with an emphasis on ethics, balance, and the undeniable human spirit that makes soccer the beautiful game.

Appendix A:
Appendix

The appendix serves as a vital tool to complement the main content of this book. Within this section, you'll find supplementary material that can deepen your understanding of the intersections between artificial intelligence and soccer. The goal is to provide additional resources, data, and references that support the topics discussed in the preceding chapters.

1. Data Tables

Here, you will find comprehensive tables containing key statistics and datasets referenced throughout the book. These tables are designed to allow readers to delve deeper into specific aspects such as player performance metrics, AI algorithms, and game analysis data.

2. Algorithm Descriptions

This section offers detailed descriptions and pseudo-code for some of the AI algorithms discussed. Feel free to explore these to grasp how these algorithms function and their practical applications in soccer.

3. Case Study Details

While Chapter 10 provided an overview of various case studies, this section expands on those by providing in-depth reports and background information. Learn how top clubs and organizations successfully implement AI technology in their operations.

4. Interviews and Surveys

Transcripts and summaries of interviews with key experts from Chapters 11 provide firsthand insights into the minds of those at the forefront of AI and soccer. Additionally, survey data that informed some of the conclusions drawn in the book are made available for further examination.

5. Additional Resources

This section lists articles, books, websites, and other material for those who want to continue exploring the fascinating and rapidly-evolving world of AI in soccer. From scholarly papers to popular articles, find resources tailored to varying levels of expertise.

6. Glossary Extensions

While the main glossary covers essential terms, this section includes additional technical jargon and soccer-specific terminology that you might encounter in specialized literature and discussions surrounding AI in soccer.

7. Ethical Guidelines and Frameworks

Further to Chapter 12, this part consolidates various ethical guidelines and frameworks that are relevant to the responsible implementation of AI in soccer. Understanding these can help stakeholders navigate the complex moral landscape presented by technology in sports.

8. Tools and Platforms

Explore various AI tools and platforms that are used in soccer analytics. This section provides summaries, capabilities, and use-case scenarios to aid in familiarizing with the tools that are shaping the future of soccer.

9. Implementation Checklists

This practical segment offers checklists and flowcharts to guide clubs and organizations in adopting and integrating AI technologies. Use these tools to ensure a smooth implementation process, aligning with best practices outlined in the book.

We believe that this appendix serves as a bridge between theoretical understanding and practical application. Its purpose is to empower you with the tools and knowledge needed to leverage AI effectively within the beautiful game of soccer.

Glossary

This glossary serves as a quick reference to key terms and concepts that are central to understanding the intersection of artificial intelligence and soccer. It is designed to help readers familiarize themselves with the terminologies used throughout the book and provide a solid foundation for further exploration of the topics discussed.

Algorithm: A set of rules or steps followed by a computer to perform a specific task. In soccer, algorithms can analyze player performance or predict game outcomes.

Artificial Intelligence (AI): A branch of computer science focused on creating systems capable of performing tasks that typically require human intelligence, such as learning, problem-solving, and decision-making.

Big Data: Extremely large datasets that can be analyzed computationally to reveal patterns, trends, and associations, especially relating to human behavior and interactions. Used extensively in soccer for performance analysis and strategic planning.

Chatbot: A software application designed to conduct conversations with users, typically over the internet. In soccer, chatbots can engage fans by providing information and answering questions in real-time.

Data-Driven Decision Making: The process of making decisions based on data analysis rather than intuition or observation alone. This approach is increasingly prevalent in soccer management and coaching.

Injury Prediction and Prevention: The use of AI and data analytics to predict the likelihood of injuries and take preventive measures to avoid them, thereby enhancing player longevity and team performance.

Machine Learning: A subset of AI that involves the use of algorithms and statistical models to enable computers to improve their performance on a task through experience. Applied in player scouting and performance analysis in soccer.

Predictive Performance Metrics: Metrics derived from AI and statistical models that forecast future player performance and potential, aiding in scouting and recruitment decisions.

Real-Time Data Analysis: The process of analyzing data as it is collected, without delay. In soccer, this can be used for immediate tactical adjustments during a game.

Virtual Coaching Assistants: AI-powered systems that provide real-time feedback and personalized training advice to players, enhancing their development and performance.

Wearable Technology: Devices such as smartwatches and fitness trackers that collect data on player movements, physical condition, and performance during training and matches.

This glossary will be updated as new technologies and terms arise, ensuring that readers stay current with the latest advancements in AI and soccer.

Resources for Further Reading

When delving deeper into the intersection of artificial intelligence and soccer, it's essential to have reliable sources that can provide both foundational knowledge and cutting-edge insights. The evolution of this dynamic field is marked by both academic research and practical implementations, which make a variety of resources invaluable.

For a comprehensive understanding of how data analytics has shaped modern soccer, consider reading "The Numbers Game: Why Everything You Know About Soccer Is Wrong" by Chris Anderson and David Sally. This book lays out the foundation of soccer analytics, effectively debunking myths and presenting how data has influenced decision-making processes in the sport. Complementing this, "Soccermatics" by David Sumpter offers a deep dive into how mathematics and data are used to decode soccer strategies and player performances, making complex concepts accessible through engaging narratives.

Understanding the fundamental principles of AI and machine learning is crucial for anyone looking to explore their applications in soccer. "Artificial Intelligence: A Guide for Thinking Humans" by Melanie Mitchell provides an insightful and approachable introduction to AI, helping readers grasp essential concepts and their implications. To focus specifically on machine learning, "Hands-On Machine Learning with Scikit-Learn and TensorFlow" by Aurélien Géron offers practical knowledge through hands-on examples, making it a must-read for tech enthusiasts and analysts aiming to apply these technologies in soccer.

Real-time data analysis and tactical adjustments are game-changers in soccer. "Statistical Modelling and Machine Learning Principles for Sports Science and Health" serves as an excellent academic resource, providing rigorous methodologies and case studies on how statistical models can revolutionize sports science. For a more practical approach,

"Football Hackers" by Christoph Biermann explores how data and AI technologies are being utilized by football clubs across Europe to enhance performance and strategy.

For those interested in player performance monitoring and the use of wearable technology, "Data-Driven Science and Engineering" by Steven L. Brunton and J. Nathan Kutz can serve as a thorough guide on data collection and its applications. This book bridges the gap between theory and practice, offering valuable insights into how wearable technology can provide crucial data for injury prediction and prevention. Additionally, the journal "Sensors" regularly features articles on the latest advancements in wearable tech, making it an excellent up-to-date resource.

To gain a better understanding of automated talent identification and predictive performance metrics, delve into "The Talent Code" by Daniel Coyle. While not exclusively focused on soccer, this book provides compelling insights into how talent can be identified and nurtured, aligning well with AI-driven scouting and recruitment methodologies. For a more technical perspective, "Machine Learning and Data Mining for Sports Analytics" compiles a series of research papers that provide deep dives into various algorithms and models used for talent identification.

In the realm of AI in training and development, "Deep Learning in Sports: From Training to Game Action" covers a range of applications from virtual coaching assistants to personalized training programs. This resource is invaluable for coaches and trainers looking to harness AI for more effective training regimes. Websites like "Coaches' Voice" also provide practical articles and videos from professional coaches, offering real-world perspectives on integrating AI into training.

Fan engagement and experience are continually transformed by AI, and "Homo Deus: A Brief History of Tomorrow" by Yuval Noah Harari discusses the looming technological shifts and their impacts on human experiences, including sports fandom. To stay current with industry trends, websites like "SportsTechie" and "SportTechie" offer frequent updates on the latest developments in fan engagement technologies.

The integration of AI in broadcasting is another rich area of exploration. "Disrupting Sports Media" by Ray Welling and Ben Green provides an academic yet accessible look at how technology, including AI, is transforming sports media and broadcasting. For enthusiasts keen on immediate applications, exploring resources from companies like Opta and Second Spectrum can reveal how AI-driven commentary systems and real-time statistics enrich the viewer experience.

For insights from case studies in AI-driven soccer, check out "Soccer Science and Performance Coaching" by Adam Owen, which includes real-life examples from top clubs that have successfully implemented AI technologies. This book not only highlights success stories but also shares lessons learned and challenges faced, offering a balanced view of AI's impact.

Insights from experts can be gleaned from "Moneyball" by Michael Lewis, which, despite its focus on baseball, offers valuable lessons on using data and analytics from a wider sports perspective. Similarly, attending conferences such as the "MIT Sloan Sports Analytics Conference" allows for networking and learning from pioneers in the field.

Ethical considerations, a significant concern with AI's increasing role in soccer, are well discussed in "Weapons of Math Destruction" by Cathy O'Neil. This book delves into the pitfalls of algorithmic decision-making, providing a critical view that is essential for understanding AI's ethical dimensions. Journals like "AI & Society" also feature articles on ethics, data privacy, and the societal impacts of AI.

Women's soccer and AI make an exciting combination, as spotlighted in "The National Team: The Inside Story of the Women Who Changed Soccer" by Caitlin Murray. This book, while not focused solely on AI, highlights the strides made in women's soccer, an area where technology like AI is starting to make significant inroads. Look for case studies published in sports science journals that focus on women's leagues for more specialized insights.

Youth development and AI is another growing field. "Outliers: The Story of Success" by Malcolm Gladwell, although not directly about soccer, provides thought-provoking ideas on talent development that align well with AI's role in identifying and nurturing young talent. Sports Academies' reports and research papers often contain detailed studies on the implementation of AI in youth programs.

Books such as "Freakonomics" by Steven D. Levitt and Stephen J. Dubner offer an intriguing perspective on data's role in making unexpected connections, akin to integrating AI with big data in soccer. For a focused read, "Data Science for Business" by Foster Provost and Tom Fawcett offers practical approaches to data integration in business scenarios, which can be extrapolated to team management applications.

To understand the financial impact of AI in soccer, "Socceronomics" by Simon Kuper and Stefan Szymanski is invaluable. The book covers the economic aspects of soccer, providing insights that can be extended to understanding AI's financial implications. For detailed financial analyses, industry reports from organizations like Deloitte on sports finances offer up-to-date, relevant data.

For a view on AI's role in player contract negotiations, "The Art of Negotiation" by Michael Wheeler provides strategies that are increasingly augmented by data-driven insights, essential for modern negotiations. Online legal and sports journals frequently publish articles on how AI is reshaping contract negotiations, providing current and practical information.

The impact of AI in refereeing, particularly with advancements like VAR, is detailed in "The Future of Sports Officiating" by Graham Poll. It covers both the technology and the human element, providing a well-rounded view. Follow publications from FIFA and UEFA.